SpringerBriefs in Water Science and Technology

More information about this series at http://www.springer.com/series/11214

Xiaochang C. Wang · Chongmiao Zhang
Xiaoyan Ma · Li Luo

Water Cycle Management

A New Paradigm of Wastewater Reuse and Safety Control

Springer

Xiaochang C. Wang
School of Environmental and Municipal Engineering
Xi'an University of Architecture and Technology
Xi'an
China

Xiaoyan Ma
School of Environmental and Municipal Engineering
Xi'an University of Architecture and Technology
Xi'an
China

Chongmiao Zhang
School of Environmental and Municipal Engineering
Xi'an University of Architecture and Technology
Xi'an
China

Li Luo
School of Environmental and Municipal Engineering
Xi'an University of Architecture and Technology
Xi'an
China

ISSN 2194-7244 ISSN 2194-7252 (electronic)
SpringerBriefs in Water Science and Technology
ISBN 978-3-662-45820-4 ISBN 978-3-662-45821-1 (eBook)
DOI 10.1007/978-3-662-45821-1

Library of Congress Control Number: 2014957758

Springer Heidelberg New York Dordrecht London

Printed on acid-free paper

Springer-Verlag GmbH Berlin Heidelberg is part of Springer Science+Business Media (www.springer.com)

Preface

Water shortage is becoming increasingly serious in many countries and regions of the world with population growth and urbanization. It is no doubt that wastewater has to be regarded as a resource, and water reclamation and reuse are, in fact, practiced almost everywhere at different scales. However, although there is no longer any question on whether or not wastewater should be reused, there are definitely many questions on how to reclaim and use the water efficiently and safely. Technologically, it is possible to remove all the pollutants from the wastewater and make it suitable for any purpose of use, but in most cities there already exist conventional water and wastewater systems, which have not been designed in a way that the reclaimed water, if produced, is easy to be embraced into the existing system as a "normal" water source. The facilities for reclaimed water production, distribution, and supply are, in most cases, like "abnormal" patches on the existing system, and the efficiency of reclaimed water utilization is also very low. Even for newly developed cities and towns, the design of the water and wastewater systems usually still follow the conventional approach.

In recent years, scientists and engineers began to reconsider our modern water and wastewater systems, which originated from systems born more than one century ago, when the world population was less than one-seventh of the current population, and natural resources, including water, were regarded as plentiful. For efficient water use and maximization of the reclaimable water sources, we have to seek changes—not merely changes for technological innovation, but also changes in our ways for system design.

To reclaim and reuse water is virtually to create a new water cycle. Therefore, water cycle management becomes a topic of study in order to manage the process of water reclamation and its uses in an efficient and safe way. Speaking about the water cycle, we have to consider what is happening in the natural hydrological cycle of world scale and/or watershed scale and try to find applicable principles from nature's example. This is the original thinking on the composition of this book, which reports the authors' recent work on water reclamation and reuse, especially in the aspects of system design, management, and safety control.

This book comprises five chapters. The first chapter introduces the authors' basic consideration on a shift from the old paradigm to a new paradigm for designing the urban water and wastewater system with reclaimed water as the important resource following the concept of water cycle management. This concept is discussed in the second chapter with conceptual explanations and mathematical models. Issues related to the safety control of water reuse are discussed in the third chapter, which stresses the ecological safety control and pathogenic safety control for non-potable use of the reclaimed water. A practical case of water cycle management for maximizing the efficiency of both freshwater and reclaimed water use is reported in the fourth chapter. The final chapter gives a summary with future perspectives. Through this book, the authors wish to share their experiences with researchers and water professionals in the field of water reuse.

Through participation in the activities of the Steering Committee of Cities of the Future Program, International Water Association (IWA) and a series of the related conferences and events, the authors obtained inspirations and encouragement to write this book. The National Natural Sciences Foundation of China has supported part of the work related to this book through two key projects (Grant Nos. 50138020, 50838005).

The authors are grateful to Dr. Mawuli Dzakpasu, Postdoctoral Research Fellow, School of Environmental and Municipal Engineering, Xi'an University of Architecture and Technology, for carefully proofreading the whole book. Special thanks also go to Ms. Fei Cong for her valuable assistance in the whole process of preparing the book.

Xiaochang C. Wang
Chongmiao Zhang
Xiaoyan Ma
Li Luo

Contents

Chapter 1
Introduction

Abstract Basic consideration was given in this chapter on a shift from the old paradigm to a new paradigm for designing the urban water and wastewater system with reclaimed water as the important resource following the concept of water cycle management.

Keywords Water shortage · Water reuse · Water cycle management

We are living in a world where fresh water resources are unevenly distributed. According to the latest working paper published by the Water Resources Institute (Gassert et al. 2013), 36 countries, all in northwest Asia, the Middle East and North Africa, are at extremely high water stress level (i.e., withdrawal/available supply > 80 %). Another 32 countries are at high water stress level (withdrawal/available supply = 40–80 %). Although China, as a whole, is at medium to high water stress level (withdrawal/available supply = 20–40 %), due to uneven distribution of the available water resources (Ministry of Water Resources 2013), most of the northern provinces are, in fact, at high or extremely high water stress levels.

The worldwide problem of water shortage has much stimulated the requirement of utilizing alternative water resources, such as wastewater reuse, rainwater harvesting, and seawater/brackish water desalination. Among them, wastewater reuse is the most common measure for mitigating water shortage because wastewater is always a stable resource everywhere, irrespective of geographic, hydrogeological, and climate conditions (Asano et al. 2007). Israel and Singapore are two examples, with countrywide application of alternative water resources. In Israel, of the 1.966 billion m^3 total water supply in 2011, 59.8 % depended on fresh water sources, whereas wastewater reuse and seawater/brackish water accounted for 24.1 and 16.1 %, respectively (Israel Water Authority 2011). On the other hand, in Singapore, water supply is depending on four national taps, namely water from the local catchment, imported water (from Malaysia), treated wastewater (named NEWater), and seawater desalination. Currently, NEWater and desalinated water contribute about 30 and 10 %, respectively for the country's water supply (PUB 2013). Thus, practicing wastewater reuse has greatly supported the sustainable development of these two extremely water-deficient countries.

X.C. Wang et al., *Water Cycle Management*,
SpringerBriefs in Water Science and Technology,
DOI 10.1007/978-3-662-45821-1_1

When treated wastewater is used for water supply, its quality should meet the requirement for reuse purposes. In the Israel case, as the reclaimed water is mainly used for agricultural irrigation, secondary treatment quality is required for large-scale wastewater treatment plants (Israel Water Authority 2011), while stabilization ponds systems are used in many small-scale wastewater treatment plants with attention paid to pathogen (such as coliform bacteria) control (Juanico and Shelef 1994). In contrast to this, as the sole purpose of water reclamation is to augment drinking water sources in Singapore, the secondary effluent is further treated in the NEWater factories by sophisticated processes such as reverse osmosis and UV radiation (Qin et al. 2006) and the finished water is sent back to source water reservoirs. In more worldwide cases, treated wastewater is used for various non-potable purposes, for which standards, regulations, and guidelines have been put forward by international organizations (UNEP 2005; WHO 2006) and national agencies (USEPA 2004; Ministry of Water Resources 2006) to set criteria for safety use of the reclaimed water. In order to meet these standards, water reclamation often need two successive treatment stages, the first being a conventional biological treatment process (such as the activated sludge process), the second being a physicochemical process (such as coagulation, sedimentation, and filtration) and sometimes even employing advanced technologies such as oxidation and carbon adsorption. Apparently, the second stage of treatment is similar to what is used for drinking water purification. Therefore, a typical process for water reclamation can be regarded as a sequence of "wastewater treatment + water purification".

Such a way of reclaimed water production is, in fact, originated from the modern urban water and wastewater system with a mode of 'end-of-the-pipe' (Wilderer 2001), which is featured by withdrawing source water (surface and/or groundwater) with a quantity to meet the demand for various uses, conducting water purification to meet the highest required quality (drinking water quality), distributing the water to all users, collecting the used water (sewage) and treating it to meet the discharge regulation, and finally dumping it to a receiving water body (Wang 2007). The so-called modern water and wastewater system mode was gradually formulated after the industrial revolution and has been used almost everywhere in the world for more than one century. However, it should be pointed out that this system mode was born in the era when the world population was no more than 2 billion (less than 1/3 of the current world population), and natural resources (including water) were considered as 'plentiful' for human consumption in an unrestricted manner. In any sense, it cannot be regarded as an unchangeable mode toward the resource-restricted future. On the basis of such an urban water and wastewater system, wastewater reuse, when is required to practice, is often added to the existing system as an ugly shaped 'patch', which cannot harmonically melt into the urban water framework.

Unlike the Israel and Singapore cases where the objectives of wastewater reuse are clear for agricultural irrigation and argumentation of drinking water sources, in most cities of other countries and regions, reclaimed water is usually miscellaneously used for various non-potable purposes such as gardening, toilet flushing, fire protection, road care and maintenance, car washing, flow augmentation of urban streams, creation of recreational lakes and ponds, and so on (Asano et al. 2007). The first difficulty faced by many cities may be the quality control of the reclaimed water distributed to each user, which involves not only technological issues but also, psychological issues because the reclaimed water may always be regarded as being from wastewater. This difficulty may partially be the reason for sending NEWater with a quality sufficiently meeting the drinking water standard to a source water reservoir than supplying it directly for drinking purposes. The second problem may be the quantity regulation between reclaimed water production and its supply and consumption because the demand for non-potable water uses, unlike normal domestic water supply, may vary hourly, daily, and seasonally to very great extents. The third problem anticipated is the efficiency of reclaimed water use because people may take the reclaimed water as to be inferior to normal tap water or even priceless, so that the water may be consumed carelessly. This problem may also be part of the reason that in Singapore, there is no additional pipeline system provided for distributing NEWater to users. The construction of large scale distribution systems for reclaimed water supply may not be economically feasible in many cases (Chen and Wang 2009).

Facing the above-mentioned problems of the old paradigm of wastewater reuse, it has become necessary to consider a new paradigm toward more efficient use of reclaimed water resources and more effective safety control. As will be discussed in the following chapters, the freshwater resources on which human lives have been sustained so far are maintained by global and/or regional natural hydrological cycles (Narasimhan 2009). The modern technologies for water and wastewater treatment are, in fact, developed following natural processes to a great extent. For wastewater reuse, why can we not follow some natural principles again to seek newer innovations? In this regard, the concept of 'water cycle management' was introduced in managing water resource development in recent years (Najia and Lustig 2006; Pouget et al. 2012). By definition (Kundzewicz 2008; Amores et al. 2013), the so-called 'water cycle' describes an area where water is supplied and wastewater is discharged through engineering means and related to part of the natural water realm. The water cycle is, in fact, a metabolic system. Only when the metabolism—a set of life-sustaining actions goes on smoothly can the system be sustained in a healthy condition (Wang and Chen 2010; Huang et al. 2013). With wastewater being used as a resource in the water district, the movement of water, as well as the substances it carries, will be much altered. Consequently, Tambo (2006) and Tambo et al. (2012) have proposed the concept of 'urban water district', which is a water metabolic space within the hydrological cycle. It stressed the harmony of artificial facilities with the water environment in the urban area and draws a clear distinction between the water areas that should be conserved and those to be utilized.

Unfortunately, the conventional water and wastewater systems in cities are seldom planned and designed following the above-mentioned principle of water cycle management. The systems are often managed in a centralized manner, where all water to be distributed is purified at discrete locations, and the wastewater collected is sent to a discrete plant for treatment and discharge. Therefore, the reuse of treated wastewater by such systems often becomes difficult. Unless there are potential consumers near the wastewater treatment plant (this may not always be the case), long distance pipelines and large distribution networks have to be constructed. In contrast to this, a water and wastewater system managed in a decentralized manner has drawn wide attention (Gaulke et al. 2008; Chen and Wang 2009; Fujiwara 2012). The principle of decentralization can mainly be characterized as an independent collection system covering smaller service areas, onsite treatment and onsite reuse, and avoidance of long distance transfer of both the collected wastewater and the treated effluent (Wang 2007). The future tendency of urban water and wastewater system planning will be integrating water supply, wastewater collection and treatment, treated wastewater reuse, and urban water environment into one framework and designing the system in a manner that the best management of a water cycle can be easily performed for both efficient wastewater reuse and safety control.

It can be expected that the concept of water cycle management can lead to a transition of wastewater reuse and safety control from the old paradigm to a new paradigm. In the following chapters, concepts, technologies, and engineering practices will be described with attention paid to new understandings on system design and safety control for reclaimed water use based on a series of theoretical, technological, and pilot studies conducted in recent years.

References

Amores MJ, Meneses M, Pasqualino J, Antón A, Castells F (2013) Environmental assessment of urban water cycle on Mediterranean conditions by LCA approach. J Clean Prod 43:84–92. doi:10.1016/j.jclepro.2012.12.033

Asano T, Burton FL, Leverenz HL, Tsuchihashi R, Tchobanoglous G (2007) Water reuse: issues, technologies, and applications. Metcalf & Eddy, McGraw-Hill, New York

Chen R, Wang XC (2009) Cost–benefit evaluation of a decentralized water system for wastewater reuse and environmental protection. Water Sci Technol 59:1515–1523. doi:10.2166/wst.2009.156

Fujiwara T (2012) Concept of an innovative water management system with decentralized water reclamation and cascading material-cycle for agricultural areas. Water Sci Technol 66(6):1171–1177. doi:10.2166/wst.2012.246

Gassert F, Reig P, Luo T, Maddocks A (2013) Aqueduct country and river basin rankings: weighted aggregation of spatially distinct hydrological indicators. Working paper. Water Resources Institute, Washington, DC, November 2013. http://www.wri.org/publication/aqueduct-country-river-basin-rankings

Gaulke LS, Borgford-Parnell JL, Stensel HD (2008) A decentralized and onsite wastewater management course: bringing together global concerns and practical pedagogy. Water Sci Technol 58(7):1397–1404. doi:10.2166/wst.2008.348

Huang CL, Vause J, Ma HW, Yu CP (2013) Urban water metabolism efficiency assessment: integrated analysis of available and virtual water. Sci Total Environ 452:19–27. doi:10.1016/j.scitotenv.2013.02.044

Israel Water Authority (2011) Challenges and solutions for the water sector in Israel and the region. http://www.water.gov.il/Hebrew/ProfessionalInfoAndData/2012/02-Israel-Water-Sector-Overall-Review.pdf

Juanico M, Shelef G (1994) Design, operation and performance of stabilization reservoirs for wastewater irrigation in Israel. Wat Res 28:175–186. doi:10.1016/0043-1354(94)90132-5
Kundzewicz ZW (2008) Water cycle. In: Jorgensen SE, Fath B (eds) Encyclopedia of ecology. Elsevier, Amsterdam, pp 3724–3729
Ministry of Water Resources (2006) Standards of reclaimed water quality (SL368-2006, in Chinese). Ministry of Water Resources of People's Republic of China, Beijing
Ministry of Water Resources (2013) China bulletin of water resources 2012 (in Chinese). Ministry of Water Resources of People's Republic of China, December 2013. http://www.mwr.gov.cn/zwzc/hygb/szygb/qgszygb/201405/t20140513_560838.html
Najia F, Lustig T (2006) On-site water recycling—a total water cycle management approach. Desalination 188(1):195–202. doi:10.1016/j.desal.2005.04.117
Narasimhan TN (2009) Hydrological cycle and water budgets. In: Likens GE (ed) Encyclopedia of inland waters. Elsevier, Oxford, pp 714–720
Pouget L, Escaler I, Guiu R, Mc Ennis S, Versini PA (2012) Global change adaptation in water resources management: the water change project. Sci Total Environ 440:186–193. doi:10.1016/j.scitotenv.2012.07.036
PUB (2013) Our water, our future. PUB Singapore. http://www.pub.gov.sg/mpublications/OurWaterOurFuture/Pages/default.aspx
Qin JJ, Kekre KA, Tao G, Oo MH, Wai MN, Lee TC, Viswanath B, Seah H (2006) New option of MBR-RO process for production of NEWater from domestic sewage. J Membr Sci 272:70–77. doi:10.1016/j.memsci.2005.07.023
Tambo N (2006) Technology in the high entropy world. Water Sci Technol 53(9):1–8. doi:10.2166/wst.2006.269
Tambo N, Nakamura M, Wang XC (2012) The fractal lentic-lotic basin complex as a conceptual basis for future urban water systems. J Water Sustain 2(2):105–120. doi:10.11912/jws.2.2.105-120
UNEP (2005) Guidelines for municipal water reuse in the Mediterranean region: Mediterranean action plan. UNEP (DEC)/MED/WG.270/Inf.19, United Nations Environment Programme, New York
USEPA (2004) Guidelines for water reuse. Publication EPA625/R-04/108. Environmental Protection Agency, Washington DC
Wang XC (2007) Application of the concept of decentralization and reuse for designing a sustainable urban water system in water deficient area. In: Huber HG, Wilderer P, Paris S (eds) Water supply and sanitation for all: obligation of the water professionals for our common future. Hans Huber AG, Berching, pp 253–264
Wang XC, Chen R (2010) Water metabolism concept and its application in designing decentralized urban water systems with wastewater recycling and reuse. In: Hao XD, Novotny V, Nelson V (eds) Water infrastructure for sustainable communities. IWA Publishing, London, pp 199–211
WHO (2006) Guidelines for the safe use of wastewater, excreta and greywater. World Health Organization, Geneva
Wilderer PA (2001) Decentralized versus centralized wastewater management. In: Lens P, Zeeman G, Lettinga G (eds) Decentralized sanitation and reuse: concepts, systems and implementation. IWA Publishing, London, pp 39–54

Chapter 2
Concepts of Water Cycle Management for Water Reuse System Design

Abstract The natural hydrological cycle is not merely a circulation of water among the ocean, atmosphere, land, surface and subsurface water bodies, but also a process of water purification through natural actions. Following the natural principles, the concepts of water cycle management were proposed for designing the urban water and wastewater systems in a new manner by which freshwater supply and use, water reclamation and reuse, and urban water environment are integrated into one water cycle to maximize the efficiency of water application and environmental benefits. This implicates a shift of system design from the conventional paradigm to a new paradigm.

Keywords Hydrological cycle · Natural action · Urban water reuse system · Shift of paradigm

2.1 Natural Hydrological Cycle

2.1.1 Global Hydrological Cycle

The earth, the planet on which human beings as well as numerous other animals are living, is covered by oceans to about 70 % of its surface and with a water volume as great as about 1.35×10^9 km^3, while in the remaining 30 % of the terrestrial area, the total water volume is estimated at about 3.6×10^7 km^3, of which about 68.9 % is trapped in glaciers and permanent snow cover, about 29.9 % is groundwater, and about 0.9 % is in the forms of soil moisture, swamp water and permafrost, so that the water in lakes and rivers takes only about 0.3 %, or about 1.08×10^7 km^3 (Shiklomanov 1998). Figure 2.1 shows the composition of water on the earth graphically.

The groundwater, though with a much larger amount of storage than lakes and rivers, as shown in Fig. 2.1, may not always be accessible because a large part of it

X.C. Wang et al., *Water Cycle Management*,
SpringerBriefs in Water Science and Technology,
DOI 10.1007/978-3-662-45821-1_2

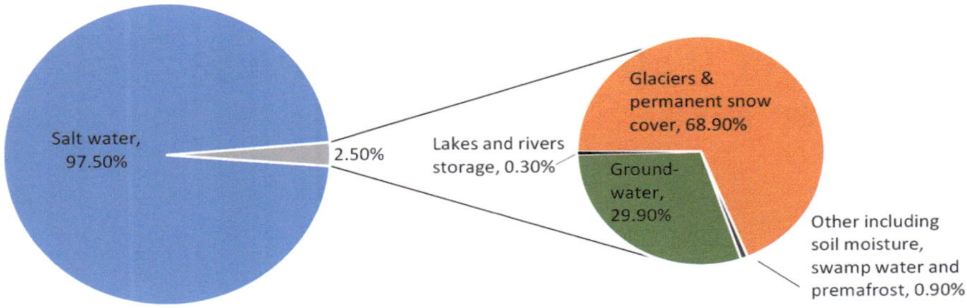

Fig. 2.1 Composition of water on the earth (graph by Wang XC)

is confined in very deep aquifers, which much decreases the exploitable groundwater quantity.

Water on the earth is not stagnant nor isolated in various water bodies but all water elements are moving, being transported from one water body to another (between ocean, atmosphere, and river/lake) and transformed from one state to another (between fluid, vapor, and ice) in a hydrological cycle as shown in Fig. 2.2. The main processes of water transport and the associated transformation include the followings (Alavian et al. 2009; Narasimhan 2009):

- Evaporation as the largest scale transport of water from the surfaces of ocean/river/lake to the atmosphere, with a transformation from the liquid state to the vapor state. The sun is providing the energy (solar energy) for the transformation and transport.

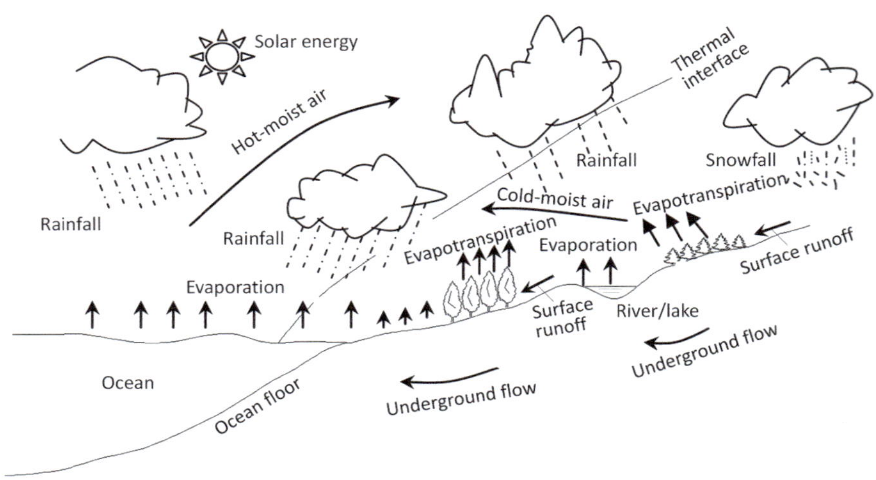

Fig. 2.2 Global hydrologic cycle (graph by Wang XC)

- Water vapor flows driven by wind power and pressure difference between wet air and dry air, which bring about a large-scale transport of the water vapor above the ocean to the terrestrial region.
- Condensation of water vapor as the moisture content in the atmosphere increases and/or temperature decreases to the saturation point, so that water elements can be transformed from the vapor state back to the liquid state again. The existence of solid particles in the atmosphere can catalyze and accelerate the condensation of water vapor.
- Precipitation of the condensed water under the action of gravitational force with water in the forms of rain, snow, and occasionally hail etc., through which water is transported back to the earth.
- Surface runoff as an important process to transport the precipitated water along a surface slope down to streams, rivers, lakes, and oceans where evaporation will take place again. It is the major final step in the closed loop of the water cycle.
- Water infiltration and groundwater flow as processes parallel with and/or during surface runoff, through which water goes down to groundwater aquifers. Under the groundwater pressure, water flows through subsurface pathways and part of it can flow back to rivers, lakes, and oceans.

The hydrological cycle is also associated with other processes such as evapotranspiration, canopy interception and so on. The annual amount of water involved in the hydrological cycle should always be dynamically balanced. This dynamic balance is a precondition for each water body on the earth to be with approximately fixed amount of water in it.

2.1.2 Hydrological Cycle of a Watershed

Down to a watershed, there still exists a hydrological cycle with similar hydrological processes as discussed above. However, as the hydrological cycle in every watershed is a subsystem of the global hydrological cycle, it is not completely independent from the global water transport and should have influence from other watersheds and the atmosphere beyond its own area (Brutsaert 2005).

As Fig. 2.3 shows, a watershed is an area confined by its dividing crest or hydrological demarcation line. It consists of both the surface part and subsurface part, though the dividing crest of the subsurface part may not always coincide with the surface part, but depend on its piestic height (for confined aquifers) or phreatic level (for unconfined or phreatic aquifers). The precipitation within the watershed area is the main source for surface and groundwater replenishment, but the origin of the precipitated water may not be limited to the amount of evaporation and evapotranspiration within its watershed area because of the global scale or trans-watersheds scale water vapor flow. Therefore, there exists a water balance relationship for a watershed as below:

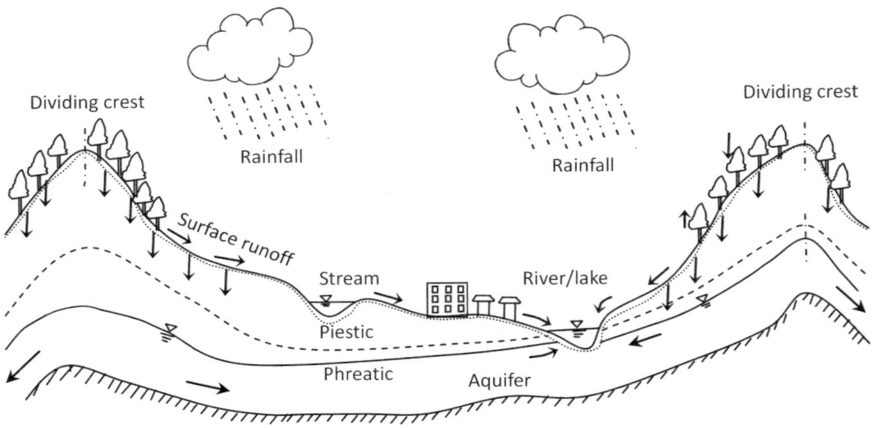

Fig. 2.3 Surface and subsurface flows in a watershed (graph by Wang XC)

$$Precipitation - Surficial\,loss - Outflow - Increase\,in\,storage = 0 \qquad (2.1)$$

where 'Precipitation' is calculated by the annual depth of rainfall and the watershed area, 'Surficial loss' is the amount of evaporation and evapotranspiration from the watershed, 'Outflow' is the amount of water flowing out of the watershed to the mainstream (for a branch river) or the ocean (for a river with ocean as the final outlet), and 'Increase in storage' includes those increased to the surface water, groundwater, soil moisture and so on. As the annual precipitation may be influenced by weather conditions outside the watershed area, the water storage in the watershed may not always be secured. This can explain why droughts occasionally occur in many watersheds in the world.

2.1.3 Functions of the Hydrological Cycle

The hydrological cycle, basically, has two important functions from the viewpoint of human needs for water resources. The first function it performs is to secure the quantity of renewable water resources on the earth's land surface under a dynamic equilibrium condition. It is estimated that every year the turnover of water on earth amounts to about 5.77×10^5 km^3. This is the water that evaporates from the oceanic surface (about 5.03×10^5 km^3) and from the land (about 7.4×10^4 km^3). The same amount of water returns back to earth as a result of precipitation, with about 4.58×10^5 km^3 falling to the ocean and about 1.19×10^5 km^3 to the land. By a simple calculation of the difference between the precipitated amount and the evaporated amount on the land, the net amount of water transported from the ocean to the land is 4.5×10^4 km^3, which provides the principal source of the renewable fresh water to support life necessities and the economic activities of human beings. Of course, an equivalent amount of water eventually discharges to the ocean through rivers (about 4.3×10^4 km^3/year) and groundwater runoff (about 2.0×10^3 km^3/year) to complete a water circulation in the global hydrological cycle (Kuchment 2004; Kundzewicz 2008a).

The second function the hydrological cycle performs is to secure the quality of the water resources. First of all, the evaporation of water from the oceanic surface is a process of water desalination by extracting only H_2O molecules from seawater while leaving salt substances in the sea. The production of such a huge amount of freshwater is powered by solar energy from the sun. In addition to this, as the precipitated water reaches the land surface, water quality is stabilized and improved at each step of the hydrological processes such as surface runoff and infiltration, and during flow and storage in each water body such as streams, rivers, lakes and groundwater aquifers, under a series of natural physical (sedimentation, entrapment etc.), physiochemical (natural coagulation, complexation and precipitation, filtration, adsorption, ion-exchange etc.), chemical (oxidation etc.), and biological (decomposition, degradation etc.) actions (Oki and Kanae 2006; Kundzewicz 2008b). Figure 2.4 schematically shows the processes possibly occurring in a river when it receives inflow of pollutants. In fact, almost all the processes developed by human beings for water and wastewater treatment can find their natural modes, or in other words, human beings have followed nature's example in developing their technologies.

Nowadays, we use the terminology of 'Natural purification' to explain the function of water quality conversion by natural processes. The most classic work

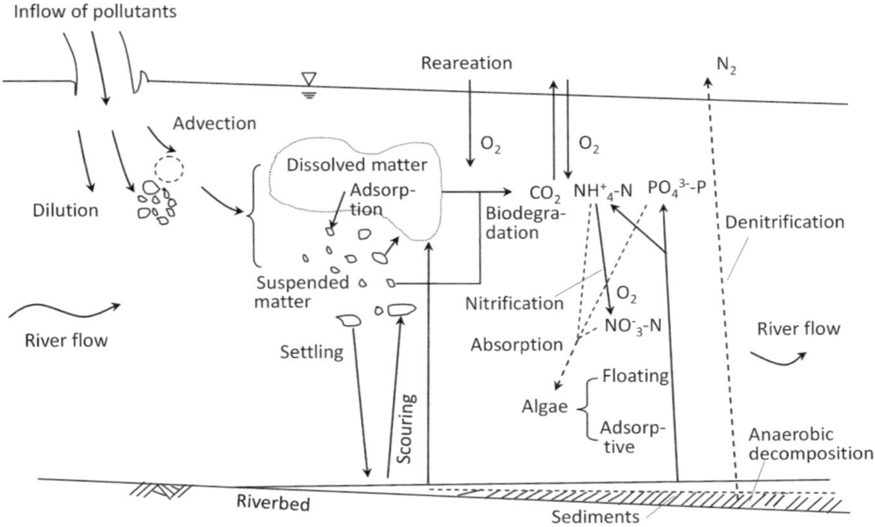

Fig. 2.4 Natural purification processes occurring in a river (graph by Wang XC)

was done by Streeter and Phelps (Ganoulis 2009) who considered both organic matter and dissolved oxygen in a stream and developed two coupled, partial differential equations as:

$$\frac{\partial C}{\partial t} + U\frac{\partial C}{\partial x} = D_T\frac{\partial^2 C}{\partial x^2} - K_1 C - K_3 C \tag{2.2}$$

$$\frac{\partial D}{\partial t} + U\frac{\partial D}{\partial x} = D_T\frac{\partial^2 D}{\partial x^2} - K_1 C - K_2 D \tag{2.3}$$

where, C: concentration of organic matter; D: oxygen deficit ($D = C_s - \mathrm{DO}$, C_s: the saturation dissolved oxygen); U: stream flow velocity; D_T: dispersion coefficient; K_1: the deoxygenation rate; K_2: the reaeration rate; K_3: the sedimentation rate. Solution of the two coupled equations gives the well-known 'oxygen sag curve' as shown in Fig. 2.5, which displays the variation of the oxygen deficit as a measure of the stream water quality.

Various studies have also been conducted for characterizing natural processes in streams, lakes, and groundwater aquifers to remove organic matter, nutrients and other pollutants (Berkun 2005; Jiang and Shen 2006).

The capability of a water body to accommodate pollutants without change in its background quality is called the 'carrying capacity' and the associated natural process within its carrying capacity is called 'self-purification.' Such a capacity that a water body possesses maintains its healthy water environmental condition as long as human disturbance is not beyond its limit of self-purification (Feng et al. 2008).

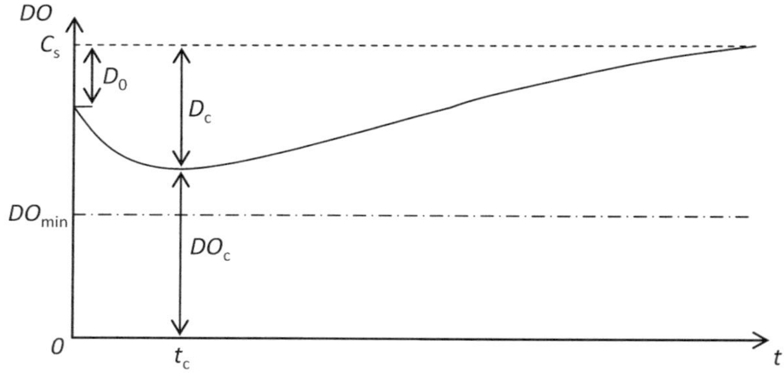

Fig. 2.5 Oxygen sag curve obtained as the solution of Streeper and Phelps equations. DO_{min} is the standardized minimum dissolved oxygen concentration; DO_c is the dissolved oxygen concentration at the *bottom* of the sag curve corresponding to time t_c; and D_0 and D_c are initial and maximum oxygen deficits, respectively (graph by Wang XC)

2.2 Urban/District Water Cycle

2.2.1 *Human Disturbance of the Hydrological Cycle*

From ancient time, people have located their homes near watercourses and used water in the simplest manner, such as taking fresh water from a nearby clean stream and then disposing of the used water arbitrarily. As the scale of water withdrawal and its final disposal was very small, there existed almost no problem on water availability and its quality deterioration because the little disturbance could be well absorbed by the sufficient carrying capacity of the water body (Feng et al. 2008). Such a condition continued from ancient time until the 1700s when the total world population was no more than 1 billion and most people were scattered in small towns and villages. Although there existed comparatively larger scale water supply by aqueduct or channel systems as earlier as 800 BC to 450 AD in Roman cities (Wolfe 1999) and AD 618–907 in the megacity of Chang'an, China at that time (Wang and Chen 2014), as these cities were all adjacent to large rivers with abundant flow, the disturbance on the local hydrological system was negligible.

Industrial revolution (later 1700s to early 1900s) was the transition point of civilization from the old era to the modern era. In addition to the transition from hand production to machines, fast urbanization and population increase were also, important symbols of revolution. Taking London as an example, its population was around 1 million in 1800, increased to nearly 2.3 million in 1850 and reached 6.5 million in 1900 (Cooper 2001). Similar urban growth also occurred in other European cities. The fast urban development brought about the birth of modern urban water systems characterized by a huge pipe network for supplying potable water to every corner of the city, and another huge sewerage network for collecting and discharging the wastewater after various water uses (Tambo 2004). As potable water is from clean source waters usually on the upstream side of the city while wastewater is finally discharged back to natural waters usually on the downstream side, a modern urban water system can be viewed as an artificial water cycle attached to natural water bodies, which are within the natural hydrological cycle (Fig. 2.6).

The extent of disturbance of urban water use on the hydrological cycle depends on the scale of the artificial water cycle shown in Fig. 2.6. For a megacity like London since early 1900s and many cities that developed later even to still larger scales, the quantity of water withdrawn from a natural water body can amount to millions m^3 per day, which substantially decreases the flow immediately downstream the water intake and disturbs the natural condition of the water body. On the other hand, the wastewater discharged back to the water body in the downstream side, no matter how it is treated, is with much inferior quality comparing with the natural water, so that water quality deterioration inevitably occurs in the water body after receiving the discharge. According to its carrying capacity, such kind of quantitative and qualitative disturbance may or may not be absorbed by the natural water bodies. From a worldwide viewpoint, as the total population was only

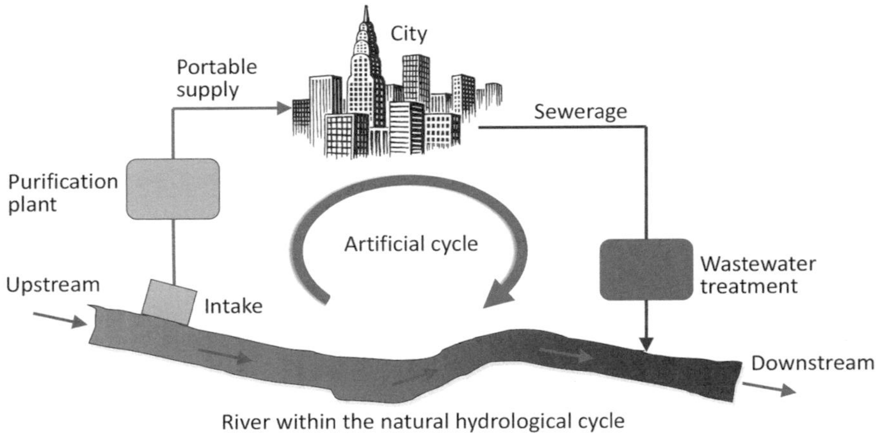

Fig. 2.6 Conventional urban water system as an artificial water cycle attached to the natural hydrological cycle (graph by Wang XC)

1 billion in 1804 and increased to 2 billion in 1927 (United Nations Population Division 2013), in the first century after the industrial revolution, there almost existed no problems in the quantity and quality of water resources, even for the watersheds where large cities were located. However, when the world population increased to 4 billion in 1974, 6 billion in 1999, and 7 billion in 2011 (United Nations Population Division 2013), centralized water withdrawal from the nature and concentrated discharge of wastewater to the nature occurred everywhere over the world and many rivers, lakes, and groundwater aquifers have been partially or completely deteriorated because of the great extent of human disturbance that has exceeded the carrying capacity of the natural hydrological cycle.

2.2.2 Conventional Urban Water System: The Old Paradigm

What we call the conventional urban water system is a system formulated more than one century ago to meet the needs of water supply and sanitation in developed cities as a result of the industrial revolution. As shown in Fig. 2.6, the conventional urban water system is characterized by centralized potable water supply under the principle of providing high quality water with sufficient quantity to meet various demands for water use, and centralized sewerage under the principle of collecting and transporting human wastes swiftly out of the urban area (Tambo 2004). In such a system, the total amount of water to be distributed to all the users over the urban area is usually withdrawn from a water source at a discrete location, and the wastewater collected in the urban area is sent to another discrete location for discharge. Therefore, huge distribution/collection networks have to be provided, and long distance water/wastewater transfer pipelines have to be constructed.

As discussed in Sect. 2.2.1, such a system was born in an era when natural water resources were thought to be plentiful. The main objective of system design was to satisfy human desire for a comfortable lifestyle and better sanitation. However, its shortcomings have become more and more obvious with the increase of world population and growth in the number and scale of cities. These include, but not limited to the followings:

- Indiscriminative use of high quality water: In almost all cities, the total quantity of water supplied to users is of potable quality, but less than half of it is used for drinking, cooking or other purposes, which really require high quality (Tambo 2004).
- Endless requirements for quality improvement: In the early 1900s when the conventional urban water system was bone, the processes for drinking water purification and wastewater treatment were 'conventional' with applications of very basic technologies. This may not be because of the lack of advanced technologies at that time, but mainly due to the easiness to obtain good quality source water and sufficient self-purification capacity of the waters to receive wastewater discharge. However, as water pollution became more and more serious in many countries and regions, especially in the vicinity of large cities, more and more sophisticated treatment was required for the provision of safe drinking water and the reduction of pollutant loading to receiving waters. High costs of water and wastewater treatment have increased economic difficulties in many cities.
- Continuous system expansion: As coverage of the whole service area is the task for the centralized urban water system, it always needs expansion and/or upgrading to meet the increasing demands due to the enlargement of urban areas and population growth. Even the maintenance and rehabilitation of the massive water and wastewater networks are hardly bearable in many cities.
- Difficulties in practicing water reuse: The conventional urban water system was designed following an 'end-of-the-pipe' model, characterized by a sequence of production-utilization-wastage (Wilderer 2001), because water reuse was not a topic at all for all the cities developed decades ago. But, when the water resource is no longer plentiful, and its reclamation and reuse become necessary, the conventional system is found to be unsuitable to meet this new requirement. Although the wastewater, which meet the quality regulation for discharge can be upgraded to a quality for reuse purposes if additional treatment is added at reasonable costs, because most of the wastewater treatment plants are located outside the city, the reclaimed water may have to be sent back to the city area by another long-distance pipeline for various purposes of reuse, unless there are potential users, such as industries and/or farm lands near the treatment plants.
- Lack of harmonic relation with natural waters: As an urban water system depends on natural waters for the provision of source water and final disposal of the discharged wastewater, it has to connect with natural waters at these two points. However, natural waters are seldom taken as components of the urban water system and the harmonic relation between engineering facilities and nature was not considered in the system design (Wang 2007).

2.2.3 Healthy Urban/District Water Cycle Design: The New Paradigm

In Sect. 2.1, we discussed the natural hydrologic cycle through which a dynamic equilibrium is maintained quantitatively and qualitatively, and various water bodies are kept 'healthy' to perform their environmental functions well. All the hydrological processes and self-purification processes in the hydrological cycle can be considered metaphorically as 'metabolism' which, by definition, is the set of chemical reactions that occur in living organisms to maintain life (Smith and Morowitz 2004). Such a metaphor was first used by Wolman (1965) in his famous paper 'The metabolism of cities' in which he developed a model to determine the inflow and outflow rates in response to deteriorating air and water qualities in American cities. Similar principles were later used for analyzing urban water (Tambo 2004) and river basins (Rodriguez-Iturbea et al. 2011). A terminological definition was given to what we call 'water metabolism' as a set of natural purification reactions to maintain a water system in a living condition (Wang and Chen 2010).

Following the concept of water metabolism, we can expand the urban water system discussed in the former sections to a larger scheme, which includes the artificial cycle of urban water use and the closely related natural waters (Fig. 2.6). Such a water system can be principally taken as an ecosystem to which the second law of thermodynamics can be applied to analyze the entropy increase as:

$$\Delta S = \oint_B \frac{\partial q}{T} \tag{2.4}$$

where ΔS is entropy increase, B is the system boundary, ∂q is any small change of energy or heat, and T is absolute temperature.

For an isolated system, it is considered to be reversible if:

$$\Delta S = 0 \tag{2.5}$$

or it is considered to be irreversible if:

$$\Delta S > 0 \tag{2.6}$$

However, since no ecosystem could ever exist as an isolated system, the second law of thermodynamics cannot be applied without adaptation. One prevailing method is to consider that the change in entropy for a non-isolated ecosystem is composed of two parts: an external contribution from outside as $\Delta_e S$ and an endogenous contribution due to the internal processes as $\Delta_i S$ (Ludovisi and Poletti 2003; Wang et al. 2011).

We may have sufficient reason to suppose that under a condition of no human disturbance at all, the natural processes occurring in the hydrological cycle can be viewed as internal processes that bring about endogenous contribution to changes in entropy, i.e. $\Delta_i S$, while the external contribution of $\Delta_e S$ can be viewed as to be from only human disturbances. Although any natural process will progress in a direction to result in entropy increase, it may still be reasonable to take the natural processes within the hydrological cycle as pseudo-reversible from its nature of self-maintenance of water and materials balance, especially when we restrict the assumption to a comparatively short time span (e.g. the time scale of human life) rather than a long time span (e.g. the time scale of natural evolution). This leads to the assumption of:

$$\Delta_i S \to 0 \tag{2.7}$$

Thermodynamically, a pseudo reversible system with ΔS approaching zero would be self-maintained so that $\Delta S \to 0$ can be a theoretical direction for healthy urban and/or district water system design to overcome the shortcomings of the conventional urban water system. Following the above discussion, as $\Delta S = \Delta_i S + \Delta_e S$ and $\Delta_i S$ can be ignored according to Eq. (2.7), there should be two directions we can follow for designing a healthy urban water system. The first direction is to decrease $\Delta_e S$, which is the entropy increase due to human disturbance on the natural hydrological cycle as far as possible, and the second direction is to make the artificial part of the urban water system as close to the natural part as far as possible so that the property of $\Delta_e S$ can be closer to that of $\Delta_i S$.

Toward the two directions, a healthy urban/district water system should be designed as a water cycle with the following characteristics:

- An enclosed water cycle with minimized supply of fresh water and minimized discharge of wastes across its boundary.
- A harmonic integration of the subsystems of water supply, sewerage, water reuse, and urban water environment in one framework.
- Treated wastewater as an important resource. As long as economically and technologically feasible, non-potable water use and environmental water use should be covered by reclaimed water.
- Introduction of natural or artificial water bodies (such as lakes, ponds, streams) into the water cycle and usage of the reclaimed water as source for water replenishment, and meanwhile for water quantity regulation and water quality polishing.

For a harmonic integration of water supply, wastewater treatment and reuse, and urban water environment, it is recommendable that a system can be structured in a form as shown in Fig. 2.7, which is called a water district with many water cycles, and using an environmental lake as a quality and quantity buffer (Tambo 2004; Tambo et al. 2012).

Recent years, with the needs for water reuse, a decentralized system has shown its advantages over the centralized system in integrating water, wastewater and reclaimed water in smaller scales, where long distance transfer pipeline and large distribution networks are unnecessary (Chen and Wang 2009; Mankad and Tapsuwan 2011; Marlow et al. 2013). Following similar principles, a decentralized system can be structured as an independent water cycle (Fig. 2.8).

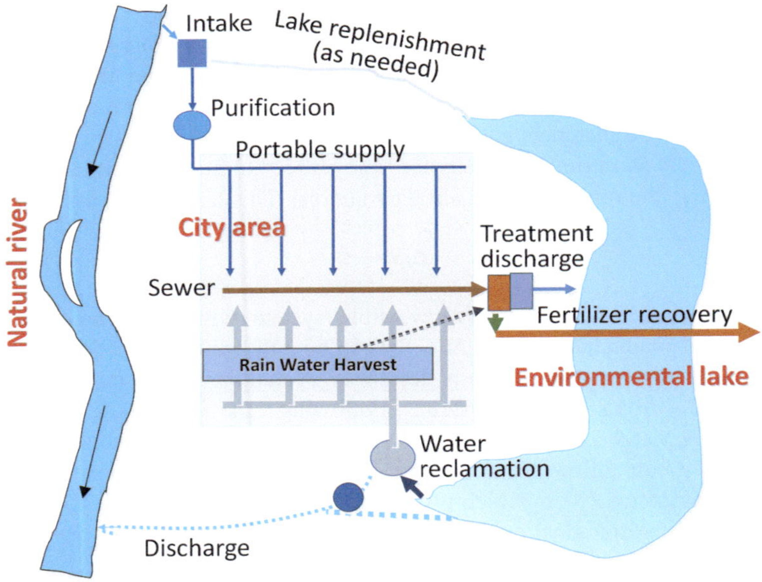

Fig. 2.7 A water district integrating water supply, wastewater treatment and reuse, and urban water environment in one framework. Adapted from Tambo et al. (2012) by permission from Journal of Water Sustainability, copyright 2012

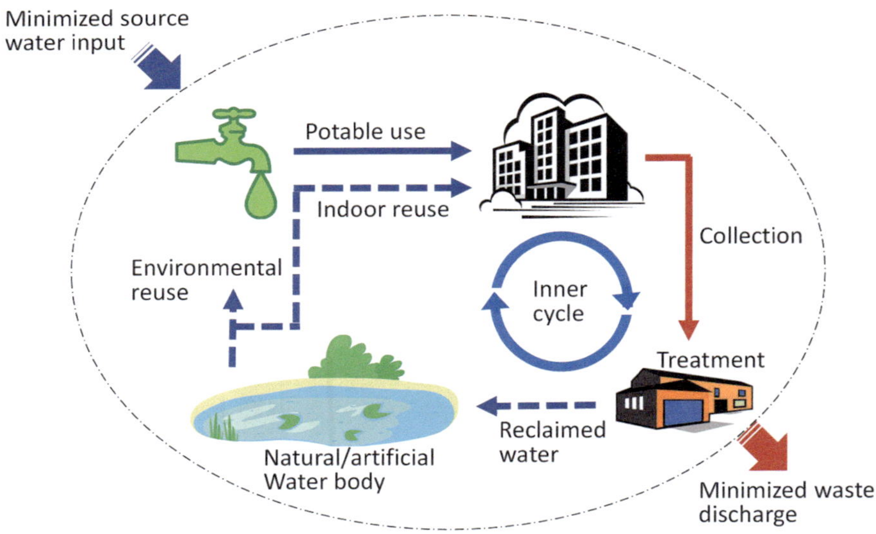

Fig. 2.8 A decentralized water system as an independent water cycle (graph by Wang XC)

2.3 Conceptual Models

2.3.1 *Models of Urban/District Water Cycle with Wastewater as a Resource*

An urban or district water system usually has four elements or subsystems, namely water resource subsystem, water supply subsystem, water use subsystem, and wastewater subsystem (Friedrich et al. 2009; Lim et al. 2010; Luo et al. 2011). Figure 2.9 outlines the relationships among these subsystems and the formation of an urban/district water cycle, which contains the artificial part (the urban/district water supply and sewerage facilities) and the natural part (the related natural waters as source water and/or receiver of urban discharge).

As wastewater is to be used as a usable source, the composition of the urban/district water system should be modified in either of the following two ways:

- Addition of an inner cycle to the urban/district water system where part of the discharged water (after quality conversion) returns to the water use subsystem to supply a certain amount of the water for certain uses that do not necessarily need freshwater. As a result, the quantity of water withdrawal from the water resource (in the water resource subsystem) and the quantity of water supply (in the subsequent water supply subsystem) are reduced. This can be called a 'water saving model' as shown in Fig. 2.10a.

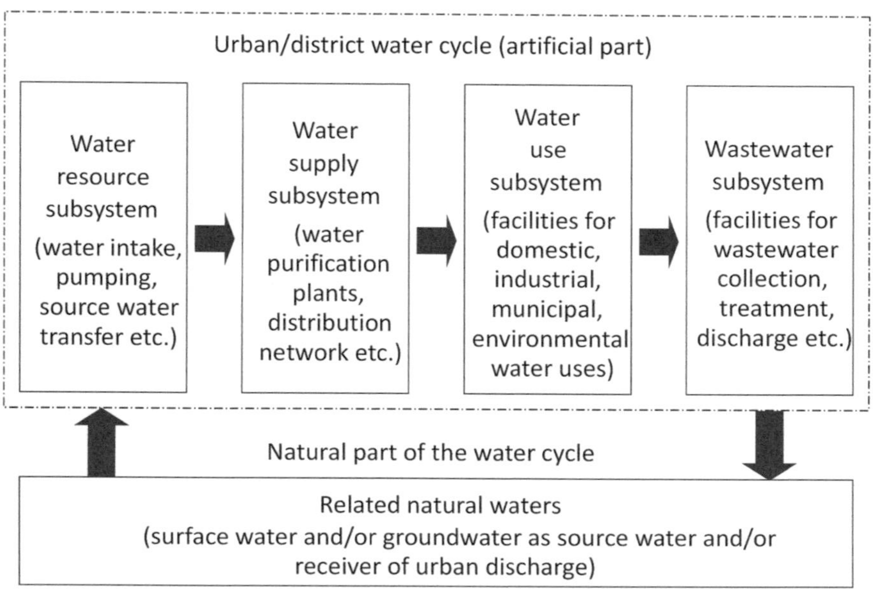

Fig. 2.9 Composition of an urban/district water system and the formation of a water cycle consisting of the artificial and natural parts (graph by Luo L)

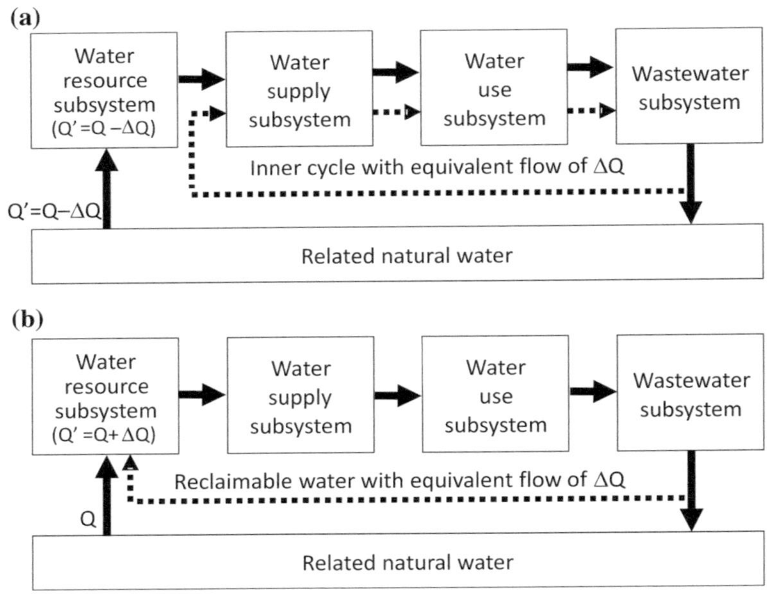

Fig. 2.10 Models of wastewater reuse for **a** water saving, and **b** water resource enlargement. ΔQ is the flow rate of the reclaimable water (graph by Wang XC and Luo L)

- An enlargement of the water source by adding an amount of the reclaimable wastewater to the source water (in the water resource subsystem). As a result, the capacity of water supply (in the water supply subsystem) to various uses (in the water use subsystem) is increased. This can then be called a 'water source enlargement model' as shown in Fig. 2.10b.

2.3.2 *Quantitative Models*

When reclaimed water is included in the urban/district water cycle as a resource, a quantitative model can be formulated (Fig. 2.11). From the viewpoint of water supply, water from either the freshwater source (surface water and/or groundwater as available) or the reclaimable water source (from the urban/district wastewater treatment and reclamation plants) will be supplied with quantities to meet the demands for potable and non-potable water uses (Bixio et al. 2006; Gikas and Tchobanoglous 2009; Willis et al. 2011) as below:

- Domestic use: household and community water uses for potable and non-potable purposes.
- Municipal use: urban public water uses, including those for public service and commercial activities of potable and non-potable purposes.

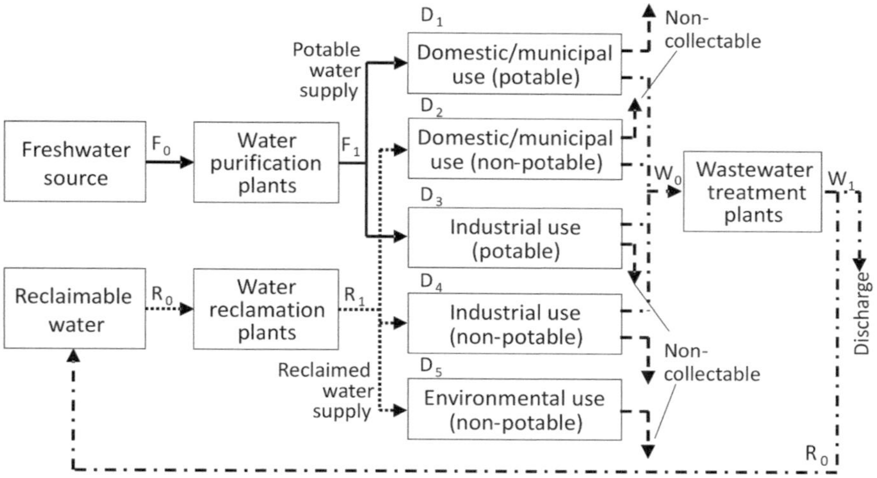

Fig. 2.11 Quantitative model of an urban/district water cycle with water reclamation and reuse. F_0: quantity of water from freshwater source; F_1: quantity of water for potable supply; D_1 and D_3: demands of potable water uses for domestic/municipal and industrial consumptions, respectively; W_0: quantity of the collected wastewater; W_1: total quantity of the treated wastewater; R_0: quantity of the reclaimable water; R_1: quantity of reclaimed water supply; D_2, D_4 and D_5: demands of reclaimed water uses for domestic/municipal, industrial, and environmental consumptions, respectively, all as m^3/d (graph by Wang XC and Luo L)

- Industrial use: water uses in industries for industrial processes and related activities of potable and non-potable purposes.
- Environmental use: water uses for environmental purposes including landscaping, gardening, urban irrigation etc., which is categorized into non-potable water consumptions.

Agricultural irrigation is not put into Fig. 2.11 because in most cases urban reuse is the main objective of water reclamation.

A water budget analysis can be conducted to analyze the relationship between various water demands and the capacity of water supply, which depends on the availability of water resources. As shown in Fig. 2.11, water demands for an urban district may mainly include those for domestic/municipal, industrial, and environmental uses. For urban water supply in the conventional manner without water reclamation and reuse, because tap water with potable water quality has to be distributed to the whole service area with sufficient quantity, irrespective of the purposes of water use, water demands are usually calculated based on population (for domestic use), population and/or service scale (for municipal use), type and production capacity of industry (for industrial use) and so on (Berger et al. 2007). However, when reclaimed water is added to the source for water supply, more detailed analysis is required on both the demands for water quantity and quality because of the qualitative difference between the reclaimed water and ordinary tap water. Table 2.1 summarizes the main water usages and corresponding requirements for potable water and/or reclaimed water as alternatives (Joksimovic et al. 2006; Berger et al. 2007).

For domestic or household water supply, reclaimed water may be preferentially used for gardening. Toilet flushing may also use reclaimed water, but this often needs the provision of dual pipe systems to supply potable and reclaimed waters by separate pipelines to each household (Jorgensen et al. 2009). For municipal water supply, reclaimed water may be preferentially used for toilet flushing in public buildings, cleaning of roads and squares, etc., urban irrigation and construction work (Willis et al. 2011). Commercial laundries, fire protection, and car washing may also be potential users of reclaimed water, but higher water quality is often required to ensure that clothes and cars may not be stained or scratched, and water quality deterioration may not easily occur in fire water tanks. Reclaimed water is most popularly used in industries for cooling and washing (Bixio et al. 2006). It can

Table 2.1 Urban water usages and corresponding requirements for water supply

Water usage	Supply requirement		Remarks[a]
	Potable water	Reclaimed water	
Domestic water use			
Drinking/cooking	××		
Bathing/washing	××		
Toilet flushing	×	×	Dual pipe needed
Gardening	×	××	
Municipal water use			
Drinking/dining services	××		
Swimming pool	××		
Commercial laundries	×	×	High quality required
Toilet flushing	×	××	
Cleaning	×	××	
Fire protection	×	×	High quality required
Urban irrigation	×	××	
Car washing	×	×	High quality required
Construction work	×	××	
Industrial water use[b]			
Cooling water	×	××	
Washing water	×	××	
Boiler water	×	×	Special quality requirement
Process water	×	×	
Environmental water use			
River flow augmentation	×	××	
Recreational lake/pond	×	××	

×× Preferential use; × Alternative use

[a] Remarks on reclaimed water use

[b] Industrial water use may also include water for drinking and/or dining that requires potable water supply

also be used as boiler water and process water, but special quality requirement often has to be taken into account (Wintgens et al. 2005).

In an urban/district water cycle with water reuse (Fig. 2.11), the freshwater from surface and/or underground sources is the primary source water with a quantity F_0, and the reclaimable water, as part of the effluent from the wastewater treatment plant, is the secondary source water with a quantity R_0. 'Secondary' here does not mean that it is less important but that it is derived from the primary source water after being used. Another feature of water flow in the water cycle (Fig. 2.11) is that the reclaimed water, after being used for certain purposes, may also add to the collectable wastewater flow, which enters the wastewater treatment plant and thus, enlarges the potential for water reclamation. Therefore, the following relationships can be established:

- Supply-demand relationship for potable water. If the freshwater source is only used for potable water supply, its quantity should meet the following relationship:

$$F_0 > F_1 \geq D_1 + D_3 \tag{2.8}$$

- Supply-demand relationship for water reclamation. If all non-potable water uses are supplied by reclaimed water, the relationship between the quantity of reclaimable water and the demands for reclaimed water supply should be as below:

$$W_1 > R_0 > R_1 \geq D_2 + D_4 + D_5 \tag{2.9}$$

- Reclaimable water source. After potable and reclaimed water uses, part of the used water can be collected through the wastewater collection system, while others are not collectable, such as those lost due to evaporation into atmosphere or infiltration into the soil, diverted out of the system, and carried away by consumers. The quantity of the reclaimable water can thus, be determined by the following relationship:

$$R_1 < R_0 \leq W_1 < W_0 = \alpha F_1 + \beta R_1 \tag{2.10}$$

  where, α is the fraction of water collected after potable water use ($\alpha < 1$), and β is the fraction of water collected after reclaimed water use ($\beta < 1$).

In the case that the wastewater treatment and water reclamation units are combined to one treatment unit as in the case of reclaimed water production directly from collected wastewater and the recovery ratio of reclaimed water production (the quantity ratio of reclaimed water to collected wastewater) is written as γ ($\gamma < 1$), then Eq. 2.10 can be rewritten as:

$$W_0 = \frac{R_1}{\gamma} = \alpha F_1 + \beta R_1 \tag{2.11}$$

The relationship between R_1, the quantity of reclaimed water supply, and F_1, the quantity of potable water supply, can then be written as:

$$\frac{R_1}{F_1} = \frac{\alpha\gamma}{1 - \beta\gamma} \tag{2.12}$$

where, $\alpha\gamma/(1 - \beta\gamma)$ can be defined as a 'recycling ratio' in an urban/district water cycle. If the reclaimed water is only used once but not recycled again, then $\beta = 0$ and the recycling ratio is $\alpha\gamma$, which is the product of α, the fraction of water collected after potable water use, and γ, the recovery ratio of reclaimed water production. The latter is mainly determined by the fraction of the collected wastewater to be used for water reclamation, in addition to the loss of water in the reclaimed water production process.

2.3.3 Materials Balance Models

In a conventional urban/district water system, water from a freshwater source is treated in a water purification plant to meet the quality requirement for potable use but the water is, in fact, supplied for both potable and non-potable uses. After use, all collectable wastewater is treated in a wastewater treatment plant to meet the quality requirement for discharging into receiving waters (Bouwer 2000; Sarkar et al. 2007). In this case, the mass balance relationship is simple because the pollutant loading in the collectable wastewater flow is the only factor to be considered in designing the wastewater treatment facilities. However, as wastewater becomes a useful resource, there is an inner water cycle within the water system as shown in Figs. 2.10 and 2.11. Therefore, in addition to the requirement of reclaimed water quality for reuse purposes, there should be a requirement on the prevention of pollutants accumulation in the water cycle. Figure 2.12 is the materials balance model of an urban/district water cycle with water reuse, for which attention is paid to the mass balance of pollutant *i* in the water cycle.

Materials balance analysis principally follows the law of conservation of mass (Chikita et al. 2004), which can be expressed as:

$$Inputs - Reacted = Outputs + Accumulated \tag{2.13}$$

For materials flow in an urban/district water cycle shown in Fig. 2.12, each of the items in Eq. 2.12 can be evaluated regarding pollutant *i* as below:

- Inputs of pollutant *i* to the water cycle: The inputs of pollutant *i* include those entering the water cycle with freshwater and reclaimed water supplies, and those added to the water cycle due to freshwater and reclaimed water uses.

$$Inputs = F_1C_{i,f} + R_1C_{i,r} + \Delta M_{i,p} + \Delta M_{i,r} \tag{2.14}$$

- Pollutant i reacted (removed): The masses of pollutant i reacted include those removed in the processes of wastewater treatment and water reclamation.

$$Reacted = \Delta M_{i,wt} + \Delta M_{i,wr} \tag{2.15}$$

- Outputs of pollutant i from the water cycle: The outputs of pollutant i are those flowing out of the water cycle by treated wastewater discharge and those contained in the waters consumed.

$$\text{Outputs} = F_{loss}C_{i,f} + R_{loss}C_{i,r} + (W_1 - R_0)\Delta M_{i,p} + \Delta M_{i,r} \tag{2.16}$$

- Accumulation of pollutant i in the water system. As shown in Fig. 2.12, because there is a closed loop of reclaimed water in the water cycle, pollutants may be accumulated in the system if the mass of pollutants returned to the treatment plant cannot be effectively removed. By substituting Eqs. 2.14–2.16 into Eq. 2.13, the accumulated pollutant mass can be calculated as:

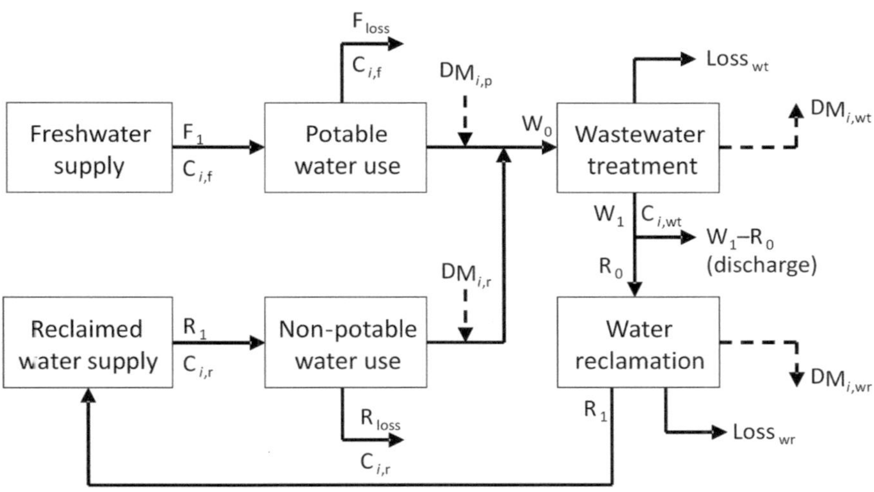

Fig. 2.12 Materials balance model of an urban/district water cycle regarding pollutant i. F_1, W_0, W_1, R_0, R_1, F_{loss}, and R_{loss}: quantities of potable water supply, collected wastewater, treated wastewater, source water for reclamation, reclaimed water supply, water lost during potable water use, and water lost during reclaimed water use, respectively, all as m^3/d; $C_{i,f}$, $C_{i,wt}$, and $C_{i,r}$: concentrations of pollutant i in potable water, treated wastewater, and reclaimed water, respectively, all as mg/L or g/m^3; F_{loss} and R_{loss}: quantities of water loss during potable and non-potable water uses, respectively, all as m^3/d; $Loss_{wt}$ and $Loss_{wr}$: quantities of water loss in wastewater treatment and water reclamation, respectively, all as m^3/d; $\Delta M_{i,p}$, $\Delta M_{i,r}$, $\Delta M_{i,wt}$, and $\Delta M_{i,wr}$: masses of pollutant i entered the water cycle or removed from the water cycle through potable water use, reclaimed water use, wastewater treatment, and water reclamation, respectively, all as g/d (graph by Wang XC and Luo L)

$$Accumulated = (F_1 - F_{loss})C_{i,f} + (R_1 - R_{loss})C_{i,r} - (W_1 - R_0)C_{i,wt} + \left(\Delta M_{i,p} + \Delta M_{i,r} - \Delta M_{i,wt} - \Delta M_{i,wr}\right) \quad (2.17)$$

- The limit condition for zero-accumulation of pollutant i in the water cycle can thus, be extrapolated from Eq. 2.17 as:

$$(F_1 - F_{loss})C_{i,f} + (R_1 - R_{loss})C_{i,r} - (W_1 - R_0)C_{i,wt} + \left(\Delta M_{i,p} + \Delta M_{i,r} - \Delta M_{i,wt} - \Delta M_{i,wr}\right) = 0 \quad (2.18)$$

References

Alavian V, Qaddumi HM, Dickson E, Diez SM, Danilenko AV, Hirji RF, Puz G, Pizarro C, Jacobsen M, Blankespoor B (2009) Water and climate change: understanding the risks and making climate-smart investment decisions. The World Bank, Washington, DC

Berger T, Birner R, Mccarthy N, DíAz J, Wittmer H (2007) Capturing the complexity of water uses and water users within a multi-agent framework. Water Resour Manage 21(1):129–148. doi:10.1007/s11269-006-9045-z

Berkun M (2005) Effects of Ni, Cr, Hg, Cu, Zn, Al on the dissolved oxygen balance of streams. Chemosphere 59:207–215. doi:10.1016/j.chemosphere.2004.11.086

Bixio D, Thoeye C, de Koning J, Joksimovic D, Savic D, Wintgens T, Melin T (2006) Wastewater reuse in Europe. Desalination 187(1):89–101. doi:10.1016/j.desal.2005.04.070

Bouwer H (2000) Integrated water management: emerging issues and challenges. Agric Water Manage 45(3):217–228. PII: S0378-3774(00)00092-5

Brutsaert W (2005) Hydrology: an introduction. Cambridge University Press, Cambridge

Chen R, Wang XC (2009) Cost–benefit evaluation of a decentralized water system for wastewater reuse and environmental protection. Water Sci Technol 59(8):1515–1522. doi:10.2166/wst.2009.156

Chikita KA, Nishi M, Fukuyama R, Hamahara K (2004) Hydrological and chemical budgets in a volcanic caldera lake: Lake Kussharo, Hokkaido, Japan. J Hydrol 291:91–114. doi:10.1016/j.jhydrol.2003.12.014

Cooper PF (2001) Historical aspects of wastewater treatment. In: Lens P, Zeeman G, Lettinga G (eds) Decentralized sanitation and reuse: concepts, systems and implementation. IWA Publishing, London, pp 11–38

Feng LH, Zhang XC, Luo GY (2008) Application of system dynamics in analyzing the carrying capacity of water resources in Yiwu City, China. Math Comput Simul 79(3):269–278. doi:10.1016/j.matcom.2007.11.018

Friedrich E, Pillay S, Buckley CA (2009) Carbon footprint analysis for increasing water supply and sanitation in South Africa: a case study. J Clean Prod 17:1–12. doi:10.1016/j.jclepro.2008.03.004

Ganoulis J (2009) Risk analysis of water pollution. Second, revised and expanded edition. Wiley-VCH Verlag GmbH & Co. KGaA, Weinherm

Gikas P, Tchobanoglous G (2009) Sustainable use of water in the Aegean Islands. J Environ Manage 90(8):2601–2611. doi:10.1016/j.jenvman.2009.01.020

Jiang JG, Shen YF (2006) Estimation of the natural purification rate of a eutrophic lake after pollutant removal. Ecol Eng 28:166–173. doi:10.1016/j.ecoleng.2006.06.002

Joksimovic D, Kubik J, Hlavinek P, Savic D, Walters G (2006) Development of an integrated simulation model for treatment and distribution of reclaimed water. Desalination 188(1):9–20. doi:10.1016/j.desal.2005.04.097

Jorgensen B, Graymore M, O'Toole K (2009) Household water use behavior: an integrated model. J Environ Manage 91(1):227–236. doi:10.1016/j.jenvman.2009.08.009

Kuchment LS (2004) The hydrological cycle and human impact on it. In: Hoekstra AY, Savenije HHG (eds) Water resources management. Eolss Publishers, Oxford

Kundzewicz ZW (2008a) Water cycle. In: Jorgensen SE, Fath B (eds) Encyclopedia of ecology. Elsevier, Oxford, pp 3724–3729

Kundzewicz ZW (2008b) Climate change impacts on the hydrological cycle. Ecohydrol Hydrobiol 8(2–4):195–203. doi:10.2478/v10104-009-0015-y

Lim SR, Suh S, Kim JH, Park HS (2010) Urban water infrastructure optimization to reduce environmental impacts and costs. J Environ Manage 91:630–637. doi:10.1016/j.jenvman.2009.09.026

Ludovisi A, Poletti A (2003) Use of thermodynamic indices as ecological indicators of the development state of lake ecosystems. 1. Entropy production indices. Ecol Model 159:203–222. doi:10.1016/S0304-3800(02)00277-6

Luo L, Wang XC, Guo WS, Ngo HH (2011) Novel conceptual models for thermodynamic analysis of urban water systems. J Water Sustain 1(2):165–175. doi:10.11912/jws.1.2.165-175

Mankad A, Tapsuwan S (2011) Review of socio-economic drivers of community acceptance and adoption of decentralised water systems. J Environ Manage 92(3):380–391. doi:10.1016/j.jenvman.2010.10.037

Marlow DR, Moglia M, Cook S, Beale DJ (2013) Towards sustainable urban water management: acritical reassessment. Water Res 47(20):7150–7161. doi:10.1016/j.watres.2013.07.046

Narasimhan TN (2009) Hydrological cycle and water budgets. In: Likens GE (ed) Encyclopedia of inland waters. Elsevier, Oxford, pp 714–720. doi:10.1016/B978-012370626-3.00010-7

Oki T, Kanae S (2006) Global hydrological cycles and world water resources. Science 313 (5790):1068–1072. doi:10.1126/science.1128845

Rodriguez-Iturbea I, Caylora KK, Rinaldo A (2011) Metabolic principles of river basin organization. Proc Natl Acad Sci USA 108(29):11751–11755. doi:10.1073/pnas.1107561108

Sarkar B, Venkateshwarlu N, Nageswara RR, Bhattacharjee C, Kale V (2007) Potable water production from pesticide contaminated surface water—a membrane based approach. Desalination 204(1):368–373. doi:10.1016/j.desal.2006.02.041

Shiklomanov IA (1998) World water resources: a new appraisal and assessment for the 21st century. UNESCO, Paris. http://webworld.unesco.org/water/ihp/publications/waterway/webpc/world_water_resources.html

Smith E, Morowitz H (2004) Universality in intermediary metabolism. Proc Natl Acad Sci USA 101(36):13168–13173. doi:10.1073/pnas.0404922101

Tambo N (2004) Urban metabolic system of water for the 21st century. Water Sci Technol: Water Supply 4(1):1–5

Tambo N, Nakamura M, Wang XC (2012) The fractal lentic-lotic basin complex as a conceptual basis for future urban water systems. J Water Sustain 2(2):105–120. doi:10.11912/jws.2.2.105-120

United Nations Population Division (2013) World population prospects: the 2012 revision. United Nations, New York

Wang XC (2007) Application of the concept of decentralization and reuse for designing a sustainable urban water system in water deficient area. In: Huber HG, Wilderer P, Paris S (eds) Water supply and sanitation for all: obligation of the water professionals for our common future. Hans Huber AG, Berching, pp 253–264

Wang XC, Chen R (2010) Water metabolism concept and its application in designing decentralized urban water systems with wastewater recycling and reuse. In: Hao XD, Novotny V, Nelson V (eds) Water infrastructure for sustainable communities. IWA Publishing, London, pp 199–211

Wang XC, Chen R (2014) Xi'an: water management and the development of city planning in history. In: Tvedt T, Oestigaard T (eds) A history of water, series III, vol 1. Water and urbanization. IB Tauris & Co Ltd, London, pp 71–88

Wang XC, Luo L, Chen R, Ngo HH (2011) Thermodynamic analysis of an urban water system with reclaimed water as supplemental water resource. Desalin Water Treat 32:307–315. doi:10.5004/dwt.2011.2715

Wilderer PA (2001) Decentralized versus centralized wastewater management. In: Lens P, Zeeman G, Lettinga G (eds) Decentralized sanitation and reuse: concepts, systems and implementation. IWA Publishing, London, pp 39–54

Willis RM, Stewart RA, Williams PR, Hacker CH, Emmonds SC, Capati G (2011) Residential potable and recycled water end uses in a dual reticulated supply system. Desalination 272(1):201–211. doi:10.1016/j.desal.2011.01.022

Wintgens T, Melin T, Schäfer A, Khan S, Muston M, Bixio D, Thoeye C (2005) The role of membrane processes in municipal wastewater reclamation and reuse. Desalination 178(1):1–11. doi:10.1016/j.desal.2004.12.014

Wolfe P (1999) History of wastewater. In: Wolfe P (ed) World of water 2000—the past, present and future. Penn Well Magazines, Tulsa, pp 24–36

Wolman A (1965) The metabolism of cities. Sci Am 213:179–190

Chapter 3
Safety Control of Reclaimed Water Use

Abstract The reclaimed water is usually used for non-potable purposes such as domestic and environmental uses. As various pollutants residual in the reclaimed water may expose negative impacts on the aquatic systems and human health, safety control of water reuse is an important research topic. This chapter discusses the pollutant source, exposure pathways, dose-response relationships, and risk assessment methods. Attentions are mainly paid to the ecological risk due to exposure to chemical pollutants and human health risk due to exposure to pathogens.

Keywords Safety control · Chemical pollutant · Pathogen · Ecological risk · Human health risk

3.1 Risks Associated with Reclaimed Water Use

3.1.1 Pollutants Possibly Existing in Reclaimed Water

The source of the reclaimed water is the wastewater mainly from domestic sewers, and sometimes mixed with industrial discharge. The pollutants contained in the wastewater often include the followings (Tchobanoglous et al. 2003; Ellis 2006; Asano et al. 2007; Rahman et al. 2009a):

- Conventional chemical substances: total suspended solids (TSS), total dissolved solids (TDS), biochemical oxygen demand (BOD), chemical oxygen demand (COD), organic nitrogen, ammonia, organic phosphorus, inorganic phosphorus etc.
- Inorganic ions: arsenic (As), boron (B), cadmium (Cd), chromium (Cr), copper (Cu), lead (Pb), mercury (Hg), nickel (Ni), selenium (Se), silver (Ag), zinc (Zn) etc.
- Trace organic chemicals: endocrine disrupting compounds (EDCs) including a variety of chemicals such as DDT, polychlorinated biphenyls (PCBs), bisphenol A (BPA), polybrominated diphenyl ethers (PBDEs), and a number of phthalates;

X.C. Wang et al., *Water Cycle Management*,
SpringerBriefs in Water Science and Technology,
DOI 10.1007/978-3-662-45821-1_3

pharmaceutical products and personal care products (PPCPs) including a variety of pharmaceuticals such as veterinary and human antibiotics, analgesics and anti-inflammatory drugs, psychiatric drugs, lipid regulators etc., personal care products such as fragrances, sun-screen agents, insect repellents etc., and antiseptics such as triclosan and chlorophene.
- Microbial pathogens: bacteria such as *E. coli*, *Salmonella*, *Shigella* etc.; viruses such as Enteroviruses, Hepatitis A and E, Noroviruses, Rotavirus, Adenovirus etc.; protozoa such as *Giardia lambilia, Cryptosporidium parvum* etc.; and helminths such as Ascaris etc.

The processes for wastewater treatment and reclamation currently being used include a series of unit operations, which are grouped together to provide various levels of treatment. These processes, namely preliminary, primary, enhanced primary, secondary, tertiary, and advanced treatment, each have specific characteristics as follows (Asano et al. 2007):

- Preliminary treatment: mainly by screening for the removal of wastewater constituents such as rags, sticks, floatables, grits, and grease.
- Primary treatment: mainly by sedimentation (and sometimes flotation) for the removal of a portion of the suspended solids and organic matter from wastewater.
- Enhanced primary treatment: enhanced removal of suspended solids and organic matter from the wastewater, typically accomplished by chemical addition or filtration.
- Secondary treatment: biological and chemical processes for the removal of suspended solids, biodegradable organic matter (in solution or suspension), and microorganisms (by disinfection, usually as the final step).
- Secondary treatment with nutrients removal: enhanced biological and chemical processes for the removal of biodegradable organics, suspended solids, and nutrients (nitrogen, phosphorus, or both).
- Tertiary treatment: additional removal of residual suspended solids (after secondary treatment), usually by granular medium filtration, surface filtration, and membranes (nutrients removal is often included).
- Advanced treatment: additional combinations of unit operations and processes for the removal of constituents that are not reduced significantly by conventional secondary and tertiary treatment for specific water reuse applications.

Although primary and/or enhanced primary treatments are still used in some countries and regions, secondary treatment (including secondary treatment with nutrients removal) has become the minimum required process before urban wastewater is discharged into receiving water bodies in most cities all over the world (Sonune and Ghate 2004). The quality of the secondary effluent can marginally meet the requirement for certain reuse purposes such as gardening and urban irrigation, but in most cases, tertiary and/or advanced treatment are required for reclaimed water production for a broader range of purposes including industrial and most domestic and municipal uses (De Koning et al. 2008). The selection of unit processes for tertiary and/or advanced treatments depends on the treatability of various pollutants in the wastewater. Table 3.1 summarizes the treatability of a spectrum of pollutants usually encountered, by a series of unit operations typically used for water reclamation (Asano et al. 2007; Tansel 2008).

Among the unit operations listed in Table 3.1, coagulation, depth filtration, surface filtration, dissolved air flotation, and disinfection belong to the category of conventional tertiary treatment processes because they are similar to the processes conventionally used for drinking water purification (Tansel 2008). By a combination of these unit operations, suspended particles, particulate organic matter, and microorganisms can be well removed. However, most of the dissolved substances (organic matter and inorganic salts), nutrients, and trace pollutants are not removable by these operations, except for coagulation, which can effectively remove phosphorus through the action of chemical precipitation (Omoike and Vanloon 1999; Sonune and Ghate 2004).

Carbon adsorption, ion exchange, electrodialysis, and advanced oxidation can be taken as unit operations for advanced treatment for water reclamation, because, as shown in Table 3.1, the main target of treatment is to remove dissolved organic matter, trace pollutants, TDS etc., which are hardly removable by conventional tertiary treatment. Of these unit operations, ion exchange and electrodialysis are mostly used in industrial water treatment for the reduction of salt content for water supply to boiler, textile, pulp and paper processes etc., which often needs a TDS concentration as low as 100 mg/L or less (Cavaco et al. 2007). For water reclamation, ion exchange and electrodialysis are also used, mainly in industries, but not for other reuse purposes. Carbon adsorption and advanced oxidation are widely used for the removal of the dissolved organic substances, especially the non-biodegradable fraction. The combination of these two processes is a standard practice, in many cases, by placing a carbon adsorption unit after oxidation for an efficient removal of the oxidized products, including trace oxidation byproducts (Ince and Apikyan 2000). When ozone is used as an oxidant, it can partially oxidize the organic molecules and improve their biodegradability (Ledakowicz et al. 2001). In this case, ozonation is often followed by biofiltration with sand or granular activated carbon as biofilm carriers. The latter is called a biological activated carbon process (Zhang et al. 1991).

Although membrane filtration is sometimes regarded as tertiary or advanced treatment, due to its unique properties for pollutant removal, all the pressure-driven membranes can be put into one group as shown in Table 3.1. It is noticeable that all the pollutants listed in the table can be removed by membranes because the cutoffs of MF, UF, NF, and RO range from micrometers to molecule sizes, irrespective of the physical, chemical, and biological properties of the pollutants in the water to be treated (Deniz et al. 2010). Therefore, direct filtration using one and/or several stages of membranes is considered to be an effective option for water reclamation (De Koning et al. 2008). Taking advantage of the physical cutoffs, the low-pressure membranes such as MF or UF can also be combined with other physicochemical operations such as coagulation and adsorption. Specific pollutants, which can easily coagulate and/or adsorb onto particle surfaces can attach to the flocs and/or adsorbents and then be removed by membrane filtration, even though the pore size

Table 3.1 Treatability of pollutants by unit operations for tertiary/advanced treatment

Pollutants	Unit operation for tertiary/advanced treatment												
	Conventional tertiary treatment					Advanced treatment				Membrane filtration			
	Coagulation	Depth filtration	Surface filtration	Dissolved air flotation	Disinfection	Carbon adsorption	Ion exchange	Electrodialysis	Advanced oxidation	Microfiltration (MF)	Ultrafiltration (UF)	Nanofiltration (NF)	Reverse osmosis (RO)
Suspended particles	×	×	×	×						×	×		
Colloidal matter	×			×				×		×	×		
Organic matter (particulate)			×	×					×	×	×	×	
Organic matter (dissolved)					×	×			×			×	×
Nitrogen							×						×
Phosphorus	×												×
Trace pollutants						×	×		×			×	×
Total dissolved solids (TDS)							×	×				×	×
Bacteria		×	×		×				×	×	×	×	×
Protozoan cysts/oocysts		×	×	×	×				×	×	×	×	×
Viruses					×				×		×	×	×

of the membrane is not fine enough for direct removal of these pollutants (Asano et al. 2007; Fujioka et al. 2012).

Another combined process for wastewater treatment and reuse is a membrane bioreactor (MBR), which combines biological treatment with an integrated membrane system to provide enhanced removal of organics and suspended solids (Alturki et al. 2010). In an MBR system, membranes (usually MF or UF) can replace the sedimentation and depth filtration for separating the biomass in the suspended liquor in the biological unit. The biomass concentration in the reactor can thus, be maintained to a level several folds higher than that in conventional activated sludge tanks, so that biological degradation can be more efficiently performed. Due to the good separation property of the membranes, solid/liquid separation can be performed efficiently as well. Therefore, the treated water is often of high quality to meet the requirement for various reuse purposes (Santasmasas et al. 2013). MBR is widely recognized as the most promising technology for water reclamation.

As the treatability of pollutants is process specific, certain pollutants may still remain in the reclaimed water if the treatment is insufficient. In many centralized and/or decentralized systems, conventional tertiary treatment in a train of coagulation, sedimentation, filtration and disinfection is applied for water reclamation from the secondary effluent. In such cases, non-biodegradable dissolved organic substances, nutrient salts, and trace organic and inorganic pollutants in the secondary effluent may pass through the tertiary treatment without sufficient removal. Microbial pathogens, such as bacteria and viruses, can be effectively inactivated by disinfection, but they can enter the reclaimed water again from other sources during transport and storage. Therefore, chemical substances and pathogens are still the two categories of pollutants existing in the reclaimed water.

3.1.2 Ecological and Human Health Risks Associated with Water Reuse

As discussed in Sect. 2.3.2, the main purposes of water reuse in the urban area include domestic or municipal uses for gardening, toilet-flushing etc., and environmental uses for river flow augmentation and replenishment of recreational lakes and ponds. Although there are successful cases of direct and indirect potable reuse, they are limited to a few countries such as Windhoek, Namibia, USA (in Los Angeles and Orange County), and Singapore (Smith 2011). In China, there is no such case at all (Yi et al. 2011). Here, we do not stress industrial applications because the risks of water reuse in industries are usually limited to special cases.

3.1.2.1 Ecological Risks

When reclaimed water is used for gardening, the ecological risks are mainly on the plants watered either by spray or surface irrigations. The plants may be contaminated by the chemicals contained in the reclaimed water due to their adsorption either on the leaves, roots or both. The chemicals, such as heavy metals, may also be accumulated in the soil and then absorbed by the plants (Xu et al. 2010; Chen et al. 2013). Upon exposure to heavy metals, the plants may be under oxidative stress, resulting in the disturbance of their metabolisms and damage to macromolecules (Hegedüs et al. 2001). The accumulation of metal ions may further lead to cellular damage, so that the cellular ionic homeostasis may be much disturbed (Hirata et al. 2005). Severe damages may occur when the actions of the accumulated heavy metals have exceeded the plants' capability of detoxification (Zenk 1996) and phytoremediation (Wahsha et al. 2012). The visible symptoms of injury on plants are leaf chlorosis, growth retardation, browning of root tips, partial necrosis, and eventually, withering and death (Mohanpuria et al. 2007).

Regarding river flow augmentation and replenishment of recreational lakes and ponds by reclaimed water, various aquatic flora and fauna species may be under risk if the reclaimed water contains toxic substances to certain levels. The aquatic and terrestrial ecosystems may be affected from a macroscopic view (Prat et al. 2013). For stagnant lakes and ponds, the higher nutrients content of the reclaimed water than natural water may result in algae growth, which considerably affects their scenic value, but not severe ecological damage (Wei et al. 2011). In contrast to this, heavy metals are noticeably toxic to aquatic organisms. The toxic effects depend not only on the kinds and concentrations of metal ions existing in the water, but also on many biotic and abiotic factors, the former including the toxicity-tolerant property, size and life stage, nutritional requirement of the organism species, while the latter include the coexisting organic substances, pH, temperature, alkalinity and hardness, inorganic ligands, sediments and so on (Wang 1987).

Biomagnification through food chains or food webs is another action among aquatic organisms after taking in toxic chemicals. The process is also called trophic transfer, with the chemicals being transferred and biomagnified from one trophic level to the next, such as in the case of the Congo River Basin where the concentration of POPs in biota was investigated (Verhaert et al. 2013). Aquatic sediments may be a long-term source of contamination to aquatic organisms, due to the deposition of various pollutants, especially to benthic animals (Kadokami et al. 2013). In the sediments, genotoxic organics can be identified from the non-polar and medium polar fractions while cytotoxic organics can be identified from the polar fraction (Vargas et al. 2001).

3.1.2.2 Human Health Risks

In the process of water reuse, even for non-potable purposes, human beings may also be under the risk of exposure to the reclaimed water through various routes, such as ingestion, inhalation and dermal contact.

If potable reuse and/or misuse of reclaimed water for drinking are not considered, the exposure by ingestion may mainly occur in recreational activities. Swimming in recreational waters receiving reclaimed water may be one case to be considered, because unaware swallowing of water is common during swimming activities. It is estimated that during active swimming for about 45 min, the average amount of water swallowed by non-adults and adults can be 37 and 16 mL, respectively (Dufour et al. 2006). Competitive swimmers may swallow up to 128 mL water (Allen et al. 1982).

Inhalation may occur more frequently when reclaimed water is used for gardening and/or dust control by water spraying. Similar effects may occur in artificial waterfalls and fountains with reclaimed water supply, as well. In such cases, aerosols are often produced as tiny water droplets suspending in the air. These droplets may be easily transported into the respiratory tracts with human breath. It has been reported that water spraying can significantly elevate the concentration of various pathogens in the air above the ambient background level for at least 200 m downwind (Camann and Moore 1988). Infections of respiratory systems may firstly happen as a result of inhalation of the contaminated air, as well as vapor, volatile substances and dust containing other harmful substances.

Direct skin contact with reclaimed water is also a potential route of exposure during its reuse. Human skin is comprised of three primary layers, namely epidermis, dermis and hypodermis. The stratum corneum is the outermost layer of the epidermis as an effective barrier to protect the human body from foreign substances. Comparing with pathogens, chemical substances may have higher skin permeability and can easily penetrate the human skin and diffuse into the human body.

3.1.3 Methods for Ecological and Human Health Risks Assessment

Of the pollutants possibly existing in the reclaimed water, as discussed in Sect. 3.1.1, those that pose risks to aquatic ecology and human health are often the organic and inorganic chemicals of trace concentrations, and pathogens. For the selection of methods for ecological and human health risks assessment, one thing to consider is the characteristics of the pollutants to cause hazards, and another thing to consider is the nature of the hazards.

As discussed in Sect. 3.1.2.1, the risk of reclaimed water use to an aquatic ecological system is due to the accumulation of pollutants, especially chemical substances, other than pathogens. These risk-related chemicals are often of low concentrations, in ppb or even ppt levels. For example, in the USEPA's Aquatic Life Criteria Table (http://water.epa.gov/scitech/swguidance/standards/criteria/current/index.cfm), 58 pollutants have been listed and mostly with criteria values for freshwater and saltwater. Although aquatic lives are thought to be under risk if the criteria are violated, their quantitative detections are usually very heavy tasks and need advanced equipment and techniques. Because the accumulation of pollutants in an aquatic system causes certain changes in the aquatic organisms and their functions, bioassays have been found to be very useful for the comprehensive assessment of the effects of pollutants using sensitive and observable organisms as bioindicators (Hansen 2007; Watson et al. 2012).

The risk of reclaimed water use to human health is due to exposure through ingestion and inhalation of, or dermal contact with the reclaimed water. Unlike aquatic systems, which if receive reclaimed water may be continuously under exposure of the toxicants, human contact with the reclaimed water is not a continuous process, unless potable reuse is the case for investigation. What has to be assessed would be if or not, and to what extent a single or short-term exposure to reclaimed water may cause a negative impact on human health. This is a process called human health risk assessment, usually consisting of four steps, namely risk identification, exposure analysis, dose-response analysis, and risk quantification (NAS 1983). For chemical pollutants, their toxic effects on humans may be acute or chronic. However, the acute infection may be rare, in the case of water reuse, due to the low concentrations of chemicals, which are definitely below the threshold of acute toxic effects. Therefore, attention is mostly paid to the chronic health effects such as that from carcinogenic chemicals. As the development of cancer is recognized as a long-term accumulation of carcinogens through continuous exposure, such as in the study for setting drinking water quality standards, in which lifetime exposure (70 years or more) is taken into account. Anyhow, this may not be the case for non-potable use of reclaimed water. In contrast to chemicals, pathogens can cause diseases immediately after infection due to single or very short-term exposure (Pond 2005; AWWA 2006; Craun 2012). For this reason, human health risk assessment for water reuse should preferably stress the exposure to pathogens.

The frameworks and methods for ecological and human health risks assessment will be introduced in detail in the following sections.

3.2 Ecological Risk Assessment and Safety Control

3.2.1 Risk Identification

As discussed in Sect. 3.1.3, for ecological risk assessment of water reuse, we pay attention mainly to chemical pollutants including inorganic and trace organic substances.

3.2.1.1 Inorganic Chemicals and Their Potential Hazards

Water contains various cationic and anionic ions, and their total mass concentration is called salinity or total dissolved solids (TDS). Similar to domestic wastewater as the source for water reclamation, the salinity of the reclaimed water is invariably higher than that of natural water (Palacios-Díaz et al. 2009; Yi et al. 2011), because almost all the unit processes listed in Table 3.1 for wastewater treatment and reclamation, except for electrodialysis and reverse osmosis, cannot remove salts. The majority of salt ions are Na^+, Cl^- and so on, which are not harmful at normal concentrations. However, when the reclaimed water with higher salt contents is used for gardening and landscape irrigation, the salts may accumulate in soils, reduce the soil functions and affect plant growth (Palacios-Díaz et al. 2009; Morugán-Coronado et al. 2011). The extent of the impact varies with many factors other than water salinity, such as soil properties and plant characteristics (Kayikcioglu 2012; Chen et al. 2015). In fact, many salt ions, such as potassium, sodium, calcium and magnesium, are necessary elements for aquatic organisms. Nevertheless, excessive intake of these elements is deleterious.

The reclaimed water may also have higher concentrations of heavy metals than natural water. Some heavy metals, such as Cd, Hg, Ni, and Pb, are harmful to aquatic organisms, even at low concentration. The predominant effects of heavy metals on aquatic organisms include the followings:

- Many heavy metals are easy to accumulate in organisms. When the concentration is accumulated to a certain degree, heavy metals can participate in the biochemical and physical reactions inside organisms, and then disturb or destroy the normal physiological function. Heavy metals accumulated in the organisms can be transferred to higher trophic levels due to predation. As a result, aquatic organisms may show chronic toxicity symptoms (Harguinteguy et al. 2014; Luo et al. 2014).
- Through biochemical reactions, heavy metals may transform into other compounds whose toxicity is much higher than the heavy metals themselves. For instance, elemental mercury can be quickly transformed into organometallic methylmercury by microorganisms in the aquatic environment. Methylmercury has the characters of accumulation, teratogenesis and strong neurotoxicity, and is difficult to be removed from the organisms (Jenkins et al. 2007).
- Many heavy metals have a high potential to associate with the –SH group of proteins, so that the activity of relevant enzymes is inhibited, and the metabolism is disturbed finally (Fulladosa et al. 2005). The impact of heavy metals on aquatic organisms may increase with a prolonged time of reclaimed water use.

3.2.1.2 Trace Organic Chemicals and Their Potential Hazards

Trace organic chemicals possibly existing in the reclaimed water are listed in Table 3.2. Of these chemicals, many are called contaminants of emerging concern (CECs), including EDCs, PPCPs, and POPs, which have been recently 'discovered' in surface waters and are suspected to be deleterious to aquatic organisms (OW/ORD Emerging Contaminants Workgroup 2008). Many CECs are not currently included in routine monitoring programs, and their presence and significance are still not clear due to a limited number of existing studies (Herrero et al. 2012; Dodder et al. 2014). CECs may enter surface waters with industrial and domestic wastewater discharges. As secondary effluent is usually the source water for water

Table 3.2 Trace organic chemicals possibly existing in reclaimed water

Categories	Example	References
Nonpolar compounds-hydrocarbons	Linear alkylbenzenes C10–C14, petroleum-derived hydrocarbons, etc.	Smital et al. (2011)
Medium-polar compounds	Coumarine, indole, phenols, etc.	Smital et al. (2011)
Surfactants	Alkylphenol polyethoxylates, linear alkylbenzene sulfonates, etc.	Smital et al. (2011)
EDCs	17-β-estradiol, estrone, ethinyl estradiol, etc.	Kasprzyk-Hordern et al. (2008), Musolff et al. (2009), Rahman et al. (2009b), Yoon et al. (2010), Sun et al. (2013), Leusch et al. (2014a, b), Tang et al. (2014)
Pharmaceuticals	Caffeine, gemfibrozil, propyphenazone, etc.	Kasprzyk-Hordern et al. (2008), Musolff et al. (2009), Rahman et al. (2009b), Yoon et al. (2010), Smital et al. (2011), Cabeza et al. (2012), Estévez et al. (2012), Ratola et al. (2012), Leusch et al. (2014b), Tang et al. (2014)
Personal care products (PPCPs)	Octocrylene, tonalide, triclosan, etc.	Kasprzyk-Hordern et al. (2008), Yoon et al. (2010), Smital et al. (2011), Cabeza et al. (2012), Ratola et al. (2012), Leusch et al. (2014b)
Dioxins	2,3,7,8-Tetrachloro-dibenzo-p-dioxin, etc.	Cabeza et al. (2012)
PAHs	Anthracene, naphthalene, pyrene, etc.	Cabeza et al. (2012), Estévez et al. (2012), Ratola et al. (2012)
POPs	Dichlorodiphenyltrichloroethane, hexachlorocyclohexanes, etc.	Ratola et al. (2012), Zhang et al. (2013a), Sharma et al. (2014)
Pesticides	Atrazine, simazine, terbutryn, etc.	Smital et al. (2011), Estévez et al. 2012, Ratola et al. (2012), Leusch et al. (2014a), Tang et al. (2014)
Disinfection by-product	Bromodichloromethane, bromoform, etc.	Leusch et al. (2014b)
Volatile organic compounds	Hepta-brominated diphenyl ether, 1,2,3-trichlorobenzene, etc.	Cabeza et al. (2012), Estévez et al. (2012), Rodriguez et al. (2012)
Flame retardants	Trichloroethyl phosphate, trichloropropyl phosphate, etc.	Cabeza et al. (2012), Estévez et al. (2012)
X-ray contrast media (XRC)	Diatrizoic acid, Iopromide, etc.	Tang et al. (2014)
[illegible]	[illegible]	[illegible]

reclamation and these CECs may not be adequately removed unless very sophisticated treatment technologies are applied, the ecological risk associated with CECs in reclaimed water is drawing wide attention.

Trace organic chemicals are those showing relatively high toxic effects on aquatic organisms at very low concentrations, such as PCBs and some pesticides, which are found to be carcinogenic, teratogenic, and mutagenic. The mechanisms for their toxic effects can be explained as toxicity caused by their chemical structures or toxicity related to the metabolic process (Zhou et al. 2004). The former is related to the transformation of their molecular structure by enzymes in the process of cell metabolism after the intake by organisms, while the latter is related to the formation of intermediate products in the process of cell metabolism where part of the intermediate products are irreversibly combined with protein or nucleic acid by covalent bonds resulting in a change of the biochemical character of the biological macromolecules. Such kind of toxicity effects can last for a long time in the organisms.

3.2.2 *Ecological Risk Assessment Tool: Bioassays*

3.2.2.1 Advantages of Bioassays for Risk Assessment

As discussed in Sect. 3.1.3, comparing with the detection of individual chemicals and safety evaluation based on water quality criteria, bioassays are found to be more useful for a comprehensive assessment of the harmful effects of the pollutants on aquatic organisms. This needs the development of sensitive and observable organisms as bio-indicator (Hansen 2007; Watson et al. 2012). The adverse effects of pollutants on the indicator organisms often deal with the aspects of lethal, growth, reproduction, morphological structure, behavior, mutagenicity and so on (Beker et al. 2013; Kokkali and van Delft 2014; Monteiro et al. 2014). Table 3.3 is a comparison of pollutant detection, usually by physicochemical and instrumental analyses, and bioassays. It can be seen that the most important feature of bioassays is that they can intuitively reflect the detrimental effects of contaminants on organisms and profoundly reveal the mechanisms of the toxic effects including the formation, development and removal. In fact, the toxicity thus evaluated regarding water samples can present the combined action of all pollutants existing in the samples. For these reasons, bioassays are widely recognized as useful tools for ecological risk assessment.

3.2.2.2 Organisms and Their Interactions in an Aquatic Ecosystem

The objective of bioassay is to assess the ecotoxicity, which is defined as the relative strength of exogenous chemicals existing in an aquatic system to contact with organisms or enter into the vulnerable parts of their bodies and cause damages

Table 3.3 Comparison of physicochemical/instrumental analyses and bioassays

Methods	Physicochemical/instrumental analyses	Bioassays
Characteristics	Only able to determine the components and concentrations of the pollutants released in the aquatic environment	Simple, high-efficient, cost-effective
	Time-consuming and difficult to qualitatively and quantitatively analyze unknown pollutants in the environmental matrix, especially new synthesized chemicals	For in vivo bioassay, the testing organisms can be easily obtained and are with local characters
	Analytical instruments are high-precision and expensive and need a strict testing environment	Characterize the comprehensive toxicity of pollutants in reclaimed water, considering their joint toxic effects
	Cannot reflect the comprehensive toxicity effect and long-term toxicity effect of pollutants on the aquatic environment	Reflect the biological effect of pollutants, including acute and chronic toxicity

References Farré and Barceló (2003), Hernando et al. (2005), Ma et al. (2014)

(Zhou et al. 2004). The aquatic organisms under the risk of damage may include plants, phytoplankton, zooplankton, lunker, tiddler, etc., as shown in Fig. 3.1, where an aquatic ecosystem is depicted.

There always exist prey and predator relationships among organisms in an aquatic ecosystem, thus, forming a food chain as shown in Fig. 3.2. Due to external supply of organics, nutrients and energy, phytoplankton (such as algae), aquatic plants (such as lemna), and microorganisms (such as bacteria) can grow. Phytoplankton becomes

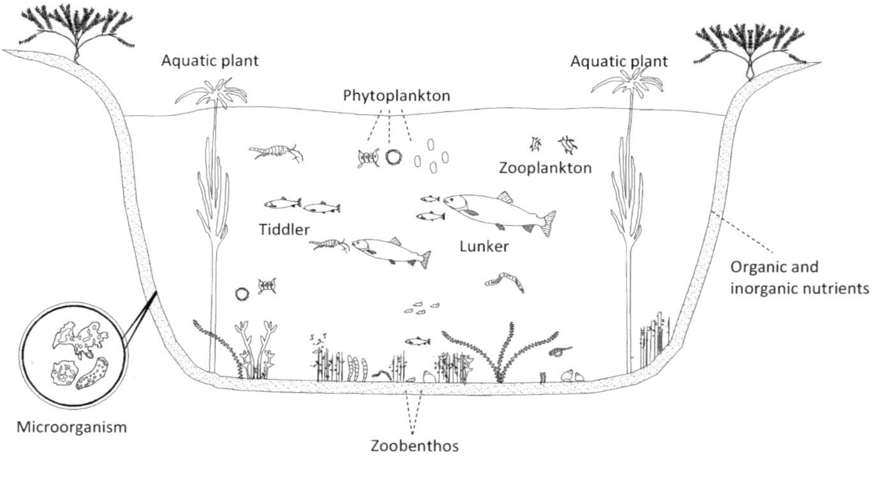

Fig. 3.1 Organisms in an aquatic ecosystem (graph by Ma XY)

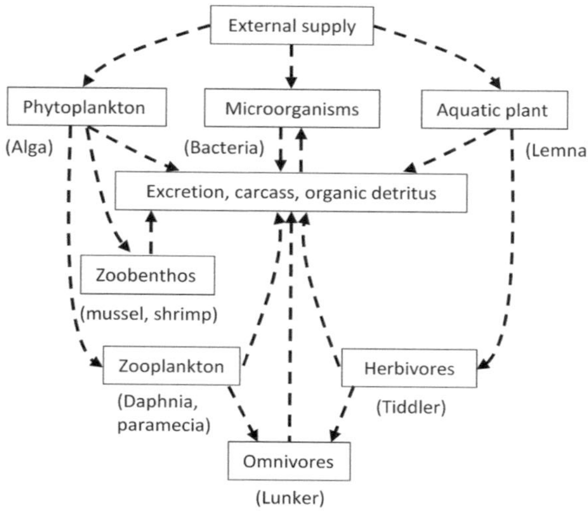

Fig. 3.2 Food chain in an aquatic ecosystem (graph by Ma XY)

food for zooplankton (such as daphnia, paramecia etc.) and zoobenthos (such as mussel, shrimp etc.), and aquatic plants become food for tiddlers as herbivores. Tiddlers, in turn, become food for carnivorous animals (such as lunker etc.) at the highest trophic level under normal conditions in the aquatic ecosystem. In the life cycles of these organisms, they discharge wastes in the form of excretes, carcasses, and detritus that can then provide organics and nutrients to microorganisms so as to form several trophic cycles. As discussed in Sect. 3.1.2.1, when toxic substances enter the aquatic system, various species of organisms may be infected (Prat et al. 2013), and biomagnification of the toxicity through the food chain may also occur (Vargas et al. 2001; Verhaert et al. 2013; Kadokami et al. 2013).

3.2.2.3 Methods of Bioassay for Ecotoxicity Assessment

For ecotoxicity assessment of aquatic ecosystems, bioassays are targeting aquatic organisms of different trophic levels as shown in Fig. 3.2. Bacteria, alga, daphnia, protozoa, and fishes are often used as testing organisms. The outlines of bioassays using each kind of aquatic organisms are described below:

- Bacterial toxicity test. The most widely used bacteria are luminescent bacteria, based on their properties of luminescence inhibition upon exposure to pollutants (Parvez et al. 2006). The species commonly used for toxicity tests include *Aliivibrio fishieri* (formerly named *Vibrio fishieri*), *Photobacterium phosphoreum*, *Vibrio harveyi* and *Vibrio qinghaiensis* sp. Q67. The bacterial toxicity is evaluated according to the extent of the impact on the visible properties of the bacteria, such as population growth, substrate consumption, respiration, and bioluminescence (Farré and Barceló 2003).

- Algal toxicity test. The commonly used algae species include *Chlorella vulgaris*, *Selenastrum capricornutum*, *Anabaena flos-aquae* and so on. As algae are primary producers in an aquatic ecosystem and play important roles in stabilizing the ecosystem through photosynthesis, they are very sensitive to the toxic action upon exposure to pollutants. The algal toxicity is evaluated according to the extent of the impact on algal growth, photosynthesis and cell abnormalities (Silva et al. 2009).
- Daphnia toxicity test. Daphnia, a freshwater invertebrate of the order Cladocera, is a key species in the aquatic food chain, because they are primary grazers of bacteria and algae, and also primary prey for fishes. Although with a small size, daphnia has almost all the internal organs with various functions similar to superior animals. Therefore, it can be used for acute and chronic toxicity tests, as well as fertility and sexuality impact tests (Guilhermino et al. 2000; Pavlaki et al. 2011). *Daphina magna* is recommended as a standard species for toxicity test in ISO 6341:1996.
- Protozoa toxicity test. Protozoa are the most primitive unicellular animals and have a close linkage with microorganisms and metazoan in the aquatic food chain. The protozoa toxicity usually reflects the impact of pollutants on its behavior, growth, morphology, submicroscopic structure and genetic damage (Takahashi et al. 2005; Hussain et al. 2008; Rico et al. 2009). *Paramecium caudatum*, *Paramecium bursaria*, and *Tetrahymena phriformis* are the commonly used species for the toxicity test.
- Fish test. Fishes are consumers in an aquatic ecosystem at the highest trophic level as shown in Fig. 3.2. They are sensitive to toxic substances and often show many symptoms after exposure to pollutants. Fishes are suitable for acute and chronic toxicity tests, and also fish accumulation test, fish embryo and larva toxicity tests (Suhendrayatna et al. 2002; Lee et al. 2011; Beker et al. 2014). The most commonly used species are *Oryzias latipes*, *Poecilia reticulate* and *Zebrafish*.

The above-mentioned bioassays are all non-specific toxicity tests on the influence of pollutants on the cell metabolism of the testing organisms. For the purpose of sophisticated toxicity evaluation, many specific toxicity tests can also be applied, such as those targeting the genotoxicity (Ames test, umu test, micronucleus test, etc.), estrogenicity (E-SCREEN assay, endocrine and androgen receptor effect, endocrine receptor-mediated effect etc.), and community structure and functioning tests (Macova et al. 2010; Leusch et al. 2014b).

3.2.2.4 Concentration-Response Relation and Toxicity Assessment

The index for quantitative assessment of toxicity are commonly selected as the median effective concentration (EC_{50}) corresponding to the response of 50 % inhibition, as well as median lethal concentration (LC_{50}) and median lethal dose (LD_{50}), which are defined as the concentration or dose causing the animal mortality

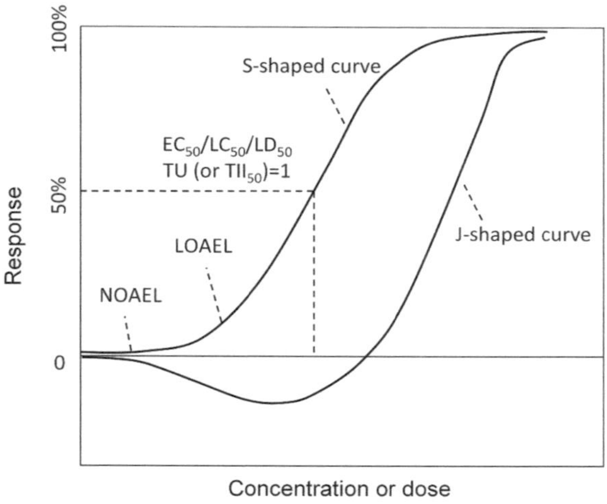

Fig. 3.3 Typical concentration/dose–response curve and the commonly used indices (graph by Ma XY)

of 50 % (Voua Otomo et al. 2014; Mori et al. 2015). These indices can be obtained from the concentration or dose–response relationship as shown in Fig. 3.3. By definition, the higher the $EC_{50}/LC_{50}/LD_{50}$ value, the lower the adverse effect is. Therefore, in most cases, a series of diluting testing samples with gradient concentrations should be prepared for the ecotoxicity tests to obtain the concentration or dose–response curve, which is usually S-shaped or J-shaped (Fig. 3.3) (Ge et al. 2011). The J-shaped curve indicates a condition of 'hormesis', which is a beneficial effect when organisms are exposed to toxicants at low concentration.

For the convenience of comparison, the toxicity units (TU) can be introduced for quantifying the toxicity of known single chemicals and mixtures (Mori et al. 2015) as:

$$TU = \frac{C}{EC_{50}} \tag{3.1}$$

where C is the concentration of the chemical. For practical water samples with unknown chemical composition, the toxicity impact index (TII_{50}) can be introduced for toxicity comparison between different waters (Farré et al. 2001) as:

$$TII_{50}(\%) = \frac{100}{EC_{50}} \tag{3.2}$$

It is also convenient to compare the toxic effect of practical water samples by selecting a known chemical as the reference substance. The toxic equivalency factor (TEF) or toxic equivalent quantity (TEQ) can then be used as toxicity indices to evaluate the single chemical and mixture toxicity, such as in the case of dioxins and dioxin-like compounds (Hong et al. 2009).

Ecotoxicity tests include acute, sub-acute or sub-chronic and chronic tests, depending on the time of exposure. The acute toxicity test is faster and more convenient to obtain results, while the chronic tests, though time consuming, is more sensitive, in terms of evaluating the toxicity of toxicants with different modes and mechanisms of action (Backhaus et al. 1997; Zhu et al. 2009). Such kind of long-term toxicity bioassay also allows a dynamic analysis of the toxicity over a certain period in the whole exposure time (Froehner et al. 2002). For sub-chronic or chronic toxicity, no observed adverse effect levels (NOAEL) and lowest observed adverse effect levels (LOAEL) can be introduced for assessing the toxicity of chemicals, so that a limited concentration as a reference value for ecological safety can be determined (Blankenship et al. 2008; Wibbertmann et al. 2011).

When the concentration of pollutants is very low in the water samples such as in the case of reclaimed water, acute toxicity is difficult to be conducted directly. For improving the sensitivity of the acute toxicity test, pretreatment has to be conducted to concentrate the target toxins before the bioassay. Solid phase extraction (SPE) is an effective method for sample pretreatment to extract the target organic substances and/or divide them into different characteristic fractions (Smital et al. 2011). Passive samplers, including semipermeable membrane devices (SPMDs) and styrenedivinylbenzene-reverse phase sulfonated (SDB-RPS) Empore™ disk, are also effective for extracting organic substances from water samples (Goodbred et al. 2009; Shaw et al. 2009). For accumulating inorganic substances, not so many pretreatment methods are available except for the method to extract metal ions from water samples by diffusive gradient in thin-films (DGTs) composed of a layer of Chelex resin impregnated on hydrogel (Roig et al. 2011). By extraction/accumulation of target substances from water samples, the interfering substances to bioassay, such as nutrients, can be simultaneously removed. However, pretreatment may inevitably be associated with the loss of certain substances that may contribute more or less to ecotoxicity. For concentrating all organic and inorganic substances from water samples, reverse osmosis is found to be useful, with effective accumulation of TOC and cationic or anionic ions, except for certain monovalent ions, such as Na^+ and Cl^- (Ma et al. 2013).

3.2.3 Bioassay for Safety Control of Water Reuse

3.2.3.1 Bioassay Using Luminescent Bacteria

Ecotoxicity assessment often needs a careful selection of testing organism in accordance with the water quality, target pollutants and objective of the assessment (Vittozzi and De Angelis 1991; Soupilas et al. 2008; Wheeler et al. 2013). This is also the case of ecotoxicity assessment for the reclaimed water. As discussed in Sect. 3.2.2, bioassay using test organisms of different trophic levels may pay attention to various aspects of the toxicity. Usually, bioassays using test organisms of a higher trophic level are more time-consuming and require skilled professionals for the test organism cultivation and manipulation. From a practical viewpoint, there is always a need for the application of fast, simple and cost-effective methods of ecotoxicity assessment. Therefore, bioassay using luminescent bacteria has become a common practice for quick toxicity screening.

Luminescent bacteria can emit yellow-green light under ordinary physiological conditions. They exist widely in the environment, such as freshwater, marine and sediments (Dunlap 2009). For ecotoxicity tests, the luminescent bacteria strains used should be nonpathogenic species, with high luminous intensity and stability, and suitable to the ambient temperature. The marine species of *Aliivibrio fishieri* (*A. fishieri*) has been identified to possess these characteristics, and is recommended as the strain for luminescent bacteria toxicity testing in ISO 11348; it is now widely used worldwide (Girotti et al. 2008).

However, bioassay based on marine *A. fischeri* needs the addition of 2–3 % salt for imitating a marine environment, which may change the properties of the testing sample. The bioavailability of metals may be reduced, and the insolubility of organic substances may be increased (Farré and Barceló 2003). As an alternative, freshwater *Vibrio qinghaiensis* sp.-Q67 (abbreviated to "Q67" hereafter) discovered in the Qinghai Province of China does not need to maintain a salt environment and has a similar mechanism of light emission with *A. fischeri*. In recent years, bioassay using Q67 has been widely applied in ecotoxicity assessment for toxicity prediction of newly synthesized compounds such as ionic liquids, and trace organic pollutants such as herbicides and pesticides (Zhang et al. 2008; Qin et al. 2011), for investigating the relationships between the toxicity and the properties of chemical compounds (Zhu et al. 2009; Wang et al. 2011), for assessing the toxicity interaction of components in the mixture (Zhang et al. 2009a, 2012a), and for assessing the toxicity of domestic wastewaters (Ma et al. 2013). A number of studies have indicated the advantage of using Q67 over marine luminescent bacteria for ecotoxicity assessment regarding the sensitivity of toxicity measurements and suitability to water quality (Liu et al. 2009a; Ye et al. 2011; Gao et al. 2012).

For ecotoxicity assessment of water reuse, because the concentration of pollutants in the water samples is usually very low, direct evaluation may become very difficult. In this case, enrichment of target substances from the water sample is necessary for expanding the detection range of the bioassay using luminescent bacteria. For example, when organic substances in water are taken as the main toxicants, solid-phase extraction (SPE) can be applied for sample pretreatment using a SPE kit equipped with suitable SPE cartridges for extracting the organic substances, followed by elution using organic solvents (Ma et al. 2011). If both organic and inorganic substances are possible toxicants, reverse osmosis (RO) is found to be useful by continuous circulation of the concentrate through a RO device (Ma et al. 2013).

3.2.3.2 Comparison of Toxicities Assessed by Different Methods

Water samples containing different organic or inorganic (metals) chemicals have been prepared for comparing their toxicities assessed by bioassays using Q67 luminescent bacteria, *C. vulgaris* as typical alga species, and *D. magna* as typical

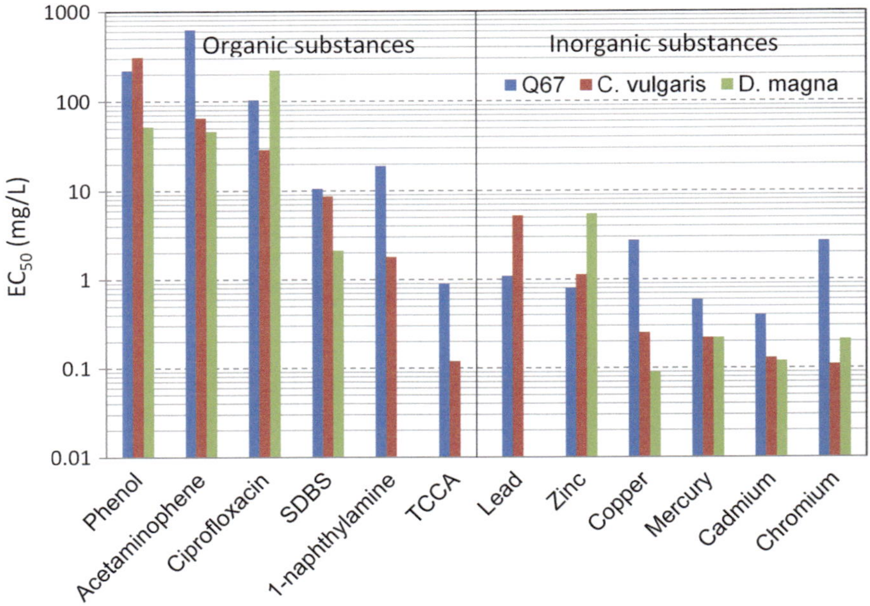

Fig. 3.4 Comparison of toxicities measured using bacteria (Q67), alga (*C. vulgaris*), and daphnia (*D. magna*). *Source* Unpublished data of the authors, graph by Ma XY

daphnia. The results are all expressed as effective concentration, EC_{50} corresponding to the mass concentration of each organic or inorganic substance to indicate the bioluminescence inhibitory effect on Q67, growth inhibition effect on *C. vulgaris*, and inhibition of mobility on *D. magna*. As shown in Fig. 3.4, three organic chemicals, namely phenol, acetaminophen, and ciprofloxacin show relatively high EC_{50} values for the three testing organisms, indicating their relatively low toxic effects on bacteria, alga, and daphnia. On the contrary, two heavy metals, namely mercury and cadmium, show relatively low EC_{50} values for the three testing organisms, indicating their relatively high toxic effects on bacteria, alga, and daphnia. Nevertheless, there are cases of much higher toxicity on alga and/or daphnia (lower EC_{50} values), but lower toxicity on bacteria (with EC_{50} values about one order higher), such as 1-naphthylamine, trichloroisocyanuric acid (TCCA), copper and chromium. Either the similarity or difference in the toxic effects of chemicals on different organisms may relate to the physicochemical property of the chemicals and the mechanisms of toxic actions. Different sensitivities of organisms to different chemicals can provide information on the selection of the testing organism according to the objective or target of toxicity assessment (Hansen 2007; Rizzo 2011).

A series of surface waters receiving different strengths of wastewater discharge have been investigated regarding their ecotoxicity levels assessed by bioassay using the Q67 luminescent bacteria and micronucleus test using *Vicia faba* root tip, which can provide information on genotoxicity (Yi and Si 2007). The Q67 acute toxicity is

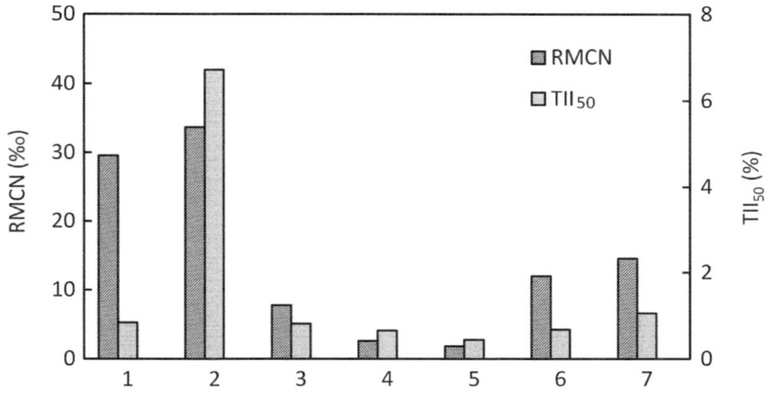

Fig. 3.5 Comparison of cytotoxicity assessed using Q67 bacteria and genotoxicity assessed using *Vicia faba* root tip. Reprinted by permission from Elsevier Limited: Ma et al. (2012), copyright 2012

expressed in term of toxicity impact index, TII_{50}, while the genotoxicity is expressed in term of the relative frequency of micronucleus, RMCN, which is the observed number of micronucleus in 1,000 cells of each *V. faba* root tip cells above the blank control. As shown in Fig. 3.5 (Ma et al. 2012), for most of the rivers monitored, higher TII_{50} occur simultaneously with higher RMCN, indicating that most of the organic substances, which are cytotoxic (toxic to the cells of bacteria) may also be genotoxic, except for sample No. 1, which is a severely polluted river.

3.2.3.3 Variation of Ecotoxicity in Wastewater Treatment and Reclamation Processes

A typical process for wastewater treatment is a secondary biological process including several stages, such as primary settling, activated sludge unit with sludge return, secondary settling, and disinfection usually by chlorination. The secondary effluent may become a water source for reclamation usually by tertiary treatment, such as coagulation followed by sand filtration. Carbon adsorption may also be applied for a reduction of soluble organic substances. Figure 3.6 gives an example of acute ecotoxicity measurement using the Q67 bacteria associated with sample pre-treatment by SEP (Ma et al. 2011). The toxicity at each stage of treatment is expressed as the toxicity impact index (TII_{50}). It can be seen that after primary settling TII_{50} is significantly reduced, while after mixing with the return sludge there is an apparent TII_{50} increase. Ecotoxicity can be effectively reduced to very low levels by biological treatment. However, as shown in the imbedded graph, a slight variation can be detected after the biological unit. Chlorination seems to bring about a certain level of increase in the ecotoxicity, possibly due to the formation of disinfection byproducts (DBPs), and coagulation shows a similar effect. As expected,

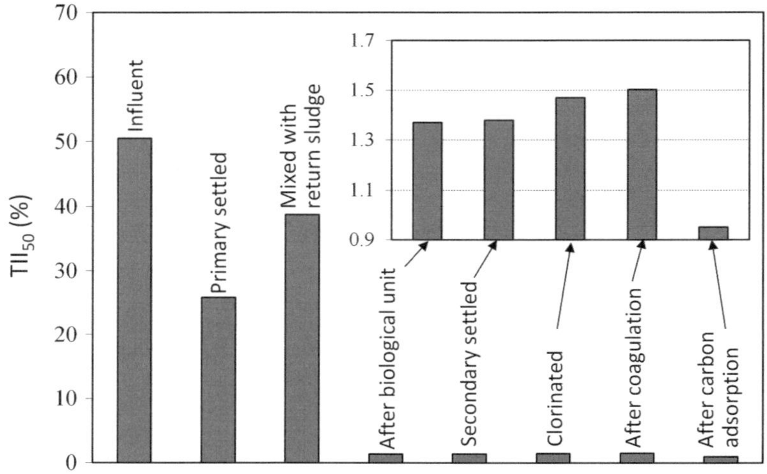

Fig. 3.6 Variation of ecotoxicity along a wastewater treatment and reclamation process. Adapted from Ma et al. (2011), by permission from Elsevier Limited, copyright 2011

carbon adsorption results in noticeable TII_{50} reduction because many soluble organic substances residual in the secondary effluent can be effectively removed.

By detailed analyses of the soluble organic substances in secondary effluent conducted by enrichment of the organic substances using an RO device and fractionation of the enriched organic substances using a series of resins to obtain six fractions based on their hydrophobicity (Wu et al. 2013), it has been identified, by three-dimensional fluorescence spectrometry analysis, that most of the organic substances existing in the secondary effluent are microbial humic-like substances, similar to those found in hydrophobic acids and hydrophilic neutrals fractions, which are THM precursors with high trihalomethane formation potential (THMFP). By a chlorination experiment and Q67 acute ecotoxicity measurements, a linear relationship is also found between THM concentrations and ecotoxicity (Fig. 3.7),

Fig. 3.7 Relationship between THM concentration and toxicity. Adapted from Wu et al. (2013), by permission from Elsevier Limited, copyright 2013

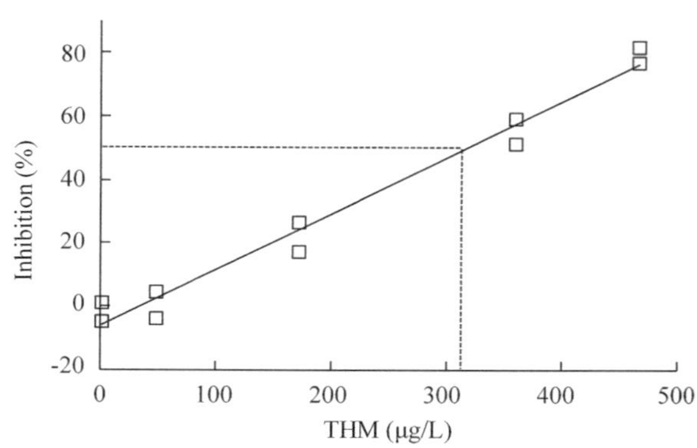

indicating that DPB formation does contribute to an increase in ecotoxicity after disinfection. An understanding of the relationship between THM formation and its potential toxicity can provide the basis for reconsidering the disinfection process for reclaimed water quality control.

Bioassay using luminescent bacteria can also be extended from acute toxicity measurements to chronic toxicity measurements by prolonging the exposure time from 5–30 min to 12–24 h, based on the changes in viability or growth rate of the bacteria (Girotti et al. 2008). Figure 3.8 is an example of chronic toxicity tests of domestic wastewater and secondary effluent using Q67 luminescent bacteria associated with RO pretreatment and the exposure time prolonged up to 12 h (Ma et al. 2013). The time-concentration-inhibition relationship can be depicted as a 3D surface showing the dynamic variation of the measured toxicity as a response to the concentration and time of exposure to toxic substances in water samples. With short exposure time, negative inhibition may occur as a result of the stimulating effect of the coexisting salts, especially for the secondary effluent sample with low pollutant concentrations, where the stimulating effect can be detected until the time of 6–7 h. However, the 3D surface becomes relatively stable after the time of 8 h, indicating a dynamic equilibrium state of the concentration–response relationship. It has also been found that the equilibrium period (8–12 h) corresponds to the logarithmic growth stage of the Q67 bacteria (Ma et al. 2013).

A combination of ecotoxicity measurements with chemical analysis can assist the identification of the origin of toxicants from various sources (Ma and Wang 2013). Bioassay using luminescent bacteria is also moving toward high automation and reproducibility with the application of advanced flow injection technology and biosensors (Komaitis et al. 2010; Lopez et al. 2012). All these have considerably broadened the area for utilizing luminescent bacteria for ecological risk assessment.

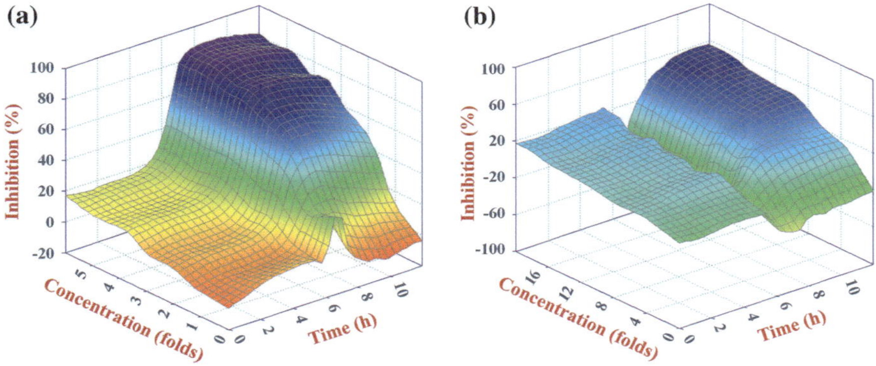

Fig. 3.8 3D inhibition surfaces for domestic wastewater (**a**) and secondary effluent (**b**). Reprinted by permission from Elsevier Limited: Ma et al. (2013), copyright 2013

3.3 Pathogenic Risk Assessment and Safety Control

3.3.1 Risk Identification: Pathogens and Their Harmful Effects on Human Health

3.3.1.1 Characteristics of Typical Pathogenic Bacteria and Viruses

Pathogenic microorganisms are critical threats to water safety all over the world. There were 780 outbreaks of diseases mainly caused by drinking water contamination in the United States from 1971 to 2006, in which the microorganism contamination took 43.9 %, while chemical pollution was only 11.5 % (Craun et al. 2010). Due to the rapid spread of many epidemic diseases, the hazards of pathogen contamination are drawing wide attention. Pathogens encountered in water, especially surface water under the influence of domestic discharge, include bacteria, viruses, parasitic protozoa and helminths. Due to their small sizes and large numbers, pathogenic bacteria and viruses are more difficult to be physically removed from water than protozoa and helminths and, therefore, become the main target of water safety control. Table 3.4 shows the major waterborne pathogens and related diseases.

Based on pathogenicity and sources, bacteria in water can be divided into two categories, namely typical pathogenic bacteria such as *Salmonella*, *Shigella*, *Campylobacter*, mostly from human and animal excreta, and opportunistic pathogenic bacteria such as *Pseudomonas*, *Legionella*, *Acinetobacter* and *Aeromonas*, mostly as indigenous bacteria in the water environment, which may not be pathogenic to normal people but fatal to those with low immunity and immunity deficiency such as infant, pregnant women, and AIDS patients.

Regarding viruses, more than 140 types are currently found in the aquatic environment, mostly as enteric viruses to cause infections in the gastrointestinal tract through fecal-oral transmission. Enteric viruses mainly include enteroviruses, rotaviruses, astroviruses, hepatitis A and E viruses, caliciviruses and enteric adenoviruses, relating to a broad spectrum of diseases ranging from fever, respiratory infections, and gastroenteritis to serious aseptic meningitis, and contributing greatly to disease morbidity and mortality across the world.

The main characteristics of the typical pathogens are as below:

- *Salmonella* spp.: *Salmonella* species belong to the family Enterobacteriaceae. They are a large group of Gram-negative bacilli composed of more than 2,000 different serotypes, but no more than three species, namely *Salmonella entericaor*, *Salmonella bongori* and *Salmonella typhi* are actually existing in water environments. *Salmonella* infection can cause a range of symptoms from mild gastroenteritis to bacteraemia or septicaemia and typhoid fever even death (WHO 2004). According to illness, *Salmonella* spp. can be classified into two fairly distinct groups, the typhoidal species, including *Salmonella typhi* and *Salmonella paratyphi* and the non-typhoidal species. The typhoid fever caused

Table 3.4 Major waterborne pathogens and related diseases

Group	Pathogen	Primary disease	Infection route
Bacteria	*Salmonella typhi*	Typhoid fever	Gastrointestinal tract
	Shigella spp.	Dysentery	Gastrointestinal tract
	Vibrio cholerae	Cholrea	Gastrointestinal tract
	Pathogenic *E. coli*	Gastroenteritis	Gastrointestinal tract
	Campylobacter jejuni	Gastroenteritis	Gastrointestinal tract
	Legionella pneumophila	Legionnaires' disease, pontiac fever	Respiratory system
	Mycobacteria spp.	Pulmonary disease	Respiratory system Skin and mucous membrane
	Aeromonas spp.	Wound infections, respiratory tract infections	Skin and mucous membrane
Viruses	Enteroviruses	Myocarditis, Meningoencephalitis, Poliomyelitis, Hand-foot-and mouth disease	Gastrointestinal tract
	Rotavirus	Acute gastroenteritis	Gastrointestinal tract
	Hepatitis A	Infectious hepatitis	Gastrointestinal tract
	Hepatitis E	Viral hepatitis	Gastrointestinal tract
	Adenovirus	Gastroenteritis	Gastrointestinal tract
	Astrovirus	Diarrhoea	Gastrointestinal tract
	Norovirus	Viral gastroenteritis	Gastrointestinal tract
Protozoa	*Cryptosporidiumparvum*	Diarrhoea	Gastrointestinal tract
	Giardia lamblia	Diarrhoea	Gastrointestinal tract
	Naegleria fowleri	Amoebic meningoencephalitis	Respiratory system
	Acanthamoeba spp.	Granulomatous amoebic encephalitis	Skin and mucous membrane
Helminths	*Dracunculus medinenis*	Urticaria, erythema, dyspnoea	Gastrointestinal tract
	Schistosoma mansoni	Schistosomiasis	Skin and mucous membrane

by typhoidal species is fatal and poses high morbidity and mortality in developing countries and with millions of cases per year reported in Asia (Crump et al. 2004). *Salmonella* spp. is the most predominant bacterial pathogen in domestic wastewater with a concentration of 10^3–10^6 copies/100 mL, mostly as non-typhoidal species (Zhang et al. 2013b). It is also detected from surface water and even drinking water (Levantesi et al. 2012).

- *Shigella* spp.: *Shigella* species are also members of the family Enterobacteriaceae, which can grow in the presence or absence of oxygen. *Shigella* is the main agent of bacillary dysentery, resulting in approximately one million deaths and

163 million cases of infection all over the world every year (Bitton 2005). *Shigella* infections always occur in regions with poor sanitation and high population density. *Shigella* can be divided into four species, namely *S. dysenteriae*, *S. flexneri*, *S. boydii* and *S. sonnei*. They can produce a toxin called Shiga toxin, which may lead to bloody diarrhea and even hemolytic uremic syndrome. The excreta from patients often contains a large number of *Shigella*, resulting in pathogenic contamination of water through sewage discharge.

- Pathogenic *Escherichia coli*: Most of *E. coli* are indigenous flora in normal intestinal tracts of humans and warm-blooded animals. Several strains of *E. coli* with virulence factors are identified as typical pathogens causing acute diarrhea, such as enterohemorrhagic *E. coli* (EHEC), enterotoxigenic *E. coli* (ETEC), enteropathogenic *E. coli* (EPEC), enteroinvasive *E. coli* (EIEC) and enteroaggregative *E. coli* (EAEC). The infectivity of EHEC, with *E. coli* O157:H7 as one serotype, is the most severe to cause bloody diarrhea with a relatively low infectious dose. *E. coli* O157:H7 causes more than 20,000 infections and as many as 250 deaths in the United States every year (Boyce et al. 1995). An outbreak of waterborne disease related to *E. coli* O157:H7 that occurred in Canada was reported to cause more than 2,300 infections and 7 deaths (WHO 2004). ETEC has a distinct feature of not invading host cells, but producing heat-stabile enterotoxin and heat-labile enterotoxin, and is an important cause of infant diarrhea mostly prevailing in tropical areas. EPEC contributes to approximately 2–8 % of *E. coli* in water (Bitton 2005), and the infections are generally associated with non-bloody diarrhea in infants and traveler's diarrhea. The symptoms of infections caused by EIEC are similar to *Shigella* as abdominal pain, watery diarrhea and fever, while EAEC can cause persistent diarrhea through aggregative adherence to host cells.
- Antibiotic-resistant bacteria: The use of antibiotics provides a condition for bacteria in wastewater to mix antibiotics continuously at sub-inhibitory concentrations (Rizzo et al. 2013), and build up their resistance to one or more classes of antibiotics (Łuczkiewicz et al. 2010). The so-called antibiotic resistant bacteria (ARB) and associated antibiotic resistant genes (ARGs) have become a global concern related to health risks. In wastewater treatment plants, the rate of ARB occurrence in total microorganisms can be higher in the treated effluent than that in the influent, and with the emerging of new ARGs (Reinthaler et al. 2003). Occurrences of multiple-antibiotic resistant pathogenic bacteria in wastewater and polluted surface waters are also documented (Walter and Vennes 1985; Mahmoud et al. 2013).
- Enteroviruses: The genus of *Enterovirus* belongs to the family Picornaviridae and consists of a single-stranded RNA genome in a non-enveloped capsid with a diameter of 20–30 nm. Members of this genus are collectively referred to as enteroviruses. Several members of enteroviruses that specifically infect humans are named human enteroviruses, including poliovirus, coxsackievirus type

A and type B, echovirus and so on. Human enteroviruses are associated with a broad spectrum of clinical manifestations ranging from mild fever to herpangina, myocarditis, meningoencephalitis, poliomyelitis, and hand-foot-mouth disease. Enteroviruses are usually excreted by infected individuals and widely distributed in the water environment. Bathing or swimming in waters contaminated by enteroviruses is a possible transmission route of waterborne diseases.

- Rotavirus: Members of the genus *Rotavirus* contain double-stranded RNA genome and double-shelled capsid with a diameter of 80 nm. Rotavirus is serologically divided into seven groups, from A to G, of which groups A–C are found in humans, especially with group A as the most important human pathogen. Rotavirus is the major cause of infantile acute gastroenteritis in children of 2 years or younger. In many developing countries in Africa, Asia and Latin America, rotavirus results in millions of children deaths every year. Rotavirus is spreading mainly by the fecal–oral route, while the respiratory route may also contribute to its transmission, such as the inhalation of aerosols containing rotavirus, in the case of reclaimed water use for artificial fountain or waterfall.
- Hepatitis A and E viruses: Hepatitis A virus (HAV) is a single-stranded RNA virus belonging to the genus *Hepatovirus* in the family Picornaviridae. The morphological and structural features of HAV are similar to enteroviruses. HAV is highly infectious to epithelial cells through the gastrointestinal tract. Infectious hepatitis can damage the liver and cause the typical symptoms of jaundice and dark urine. HAV usually spreads by the fecal-oral route, either by person-to-person direct contact or foodborne and waterborne transmission. There are many evidences that food and water contaminated by the feces of patients are important sources of the virus. In the United States, about 4 % of hepatitis cases were found to be associated with waterborne transmission in the 1970s (Cliver 1985). Hepatitis E virus (HEV) consists of a single-stranded RNA genome in a non-enveloped capsid. HEV shares properties with a number of viruses but with unidentified classification (WHO 2004). Compared with HAV, HEV transmission is mainly via fecal contaminated water rather than person-to-person (Fleisher 2012).
- Norovirus: The genus of *Norovirus*, previously called Norwalk-like virus, consists of single-stranded RNA and belong to the family Caliciviridae. There are indications of genomic variation in noroviruses recently, and 5 genogroups and 22 genetic clusters have been reported (Vinje et al. 2004). Norovirus is the common cause of acute gastroenteritis worldwide. In the United States, more than 20 million illnesses are due to infections with noroviruses, while in Finland, of the 41 outbreaks of waterborne diseases from 1998 to 2003, 18 outbreaks were confirmed to be caused by noroviruses, of which 15 outbreaks were due to the contamination of groundwater wells, whereas 3 outbreaks were associated with polluted surface waters (Maunula et al. 2005).

3.3.1.2 Methods for Concentration and Detection of Pathogens in Water Samples

Accurate and effective detection of pathogens are preconditions for further study on their distribution in water and related risk assessment. As contaminated water often contains various pathogens and other contaminants, the enrichment and purification of a specific pathogen are important prior to its detection.

Of the methods for pathogen enrichment, membrane filtration is widely used. For bacteria, due to their relatively large sizes, mechanical entrapment by membranes often plays a major role for concentrating bacteria. Regarding viruses, mechanical entrapment may only perform its role when viruses adhere to the larger solid particles while adsorption may perform the major role through various mechanisms, such as electrostatic force, hydrophobic and hydrogen bonding. A membrane adsorption-elution method can achieve a virus recovery efficiency of 73 % with carefully selected membrane pore size and at a well-controlled pH value and magnesium ion concentration (Zhang et al. 2007).

Purification or isolation of target pathogens commonly depends on the culture media, which contain substances to support the growth of specific organisms. For bacteria, culture media are nutrients such as beef extract and peptone, as well as some specific chemicals, while for viruses, due to their obligate parasitism, the culture media are host cells. However, not all microorganisms can be cultured on specific media, especially most viruses. In addition, cell culture methods are laborious, time-consuming, expensive, and requiring sophisticated and dedicated laboratories, as well as highly trained personnel. These have much limited the application of cell culture methods in routine examinations.

With the rapid development of molecular biology, polymerase chain reaction (PCR) has become a powerful tool for pathogen detection by exponentially amplifying specific DNA fragments. By PCR techniques, it is not necessary to separate target microorganisms from others, and trace amounts of specific DNA can be amplified for millions of times in a short period. Based on the measurement of fluorescence during PCR reaction, a real-time PCR technique has made it possible to quantify target microorganisms. By real-time reverse transcription-polymerase chain reaction (RT-PCR), RNA viruses can also be detected.

Unlike the medical sciences which always target certain individual serotype of viruses using specifically designed primer pairs, for health risk assessment of the water environment including reclaimed water uses, pathogen group may become the target of study, which needs simultaneous detection of viruses of a broad range of serotypes (Liu et al. 2009b). For example, respective detection of different enteroviruses may not always be required, and their collective strength can provide important information on water safety. For this purpose, universal primer pairs covering various enteroviruses have been designed according to the homology of the highly conserved 5′ noncoding region gene sequences and the simultaneous detection of enterovirus groups is realized (Zhang et al. 2010). The detection limit can be as low as 2.31 copies per liter and inter-assay variations can be lower than

5 %. The method validity has been confirmed by its application to many surface waters, wastewater and reclaimed water (Zhang and Wang 2012).

3.3.1.3 Inactivation/Removal of Pathogens in Wastewater Treatment and Reclamation

Inactivation and/or removal of pathogens are important tasks for wastewater treatment and water reclamation. For wastewater treatment, the activated sludge process still plays an important role in most countries, while for water reclamation, tertiary treatments such as sand filtration and membrane filtration are widely replied. In any sense, disinfection using chemicals is the most import process for pathogen inactivation.

In the activated sludge process, microbial flocs, called activated sludge, consist of a large number of microorganisms including predators such as ciliated protozoa. They contribute much to the reduction of bacteria in the liquid phase by grazing. In addition, adsorption and encapsulation within sludge also reduce bacteria, especially in the secondary settler. Activated sludge can achieve 0.5–2-log removal of enteric bacteria and 0.6–2-log removal of enteric viruses (Metcalf and Eddy 2007). About 1-log removal of *Salmonella* by secondary treatment is also reported (Zhang et al. 2013b).

By sand filtration, as the most common tertiary treatment process, harmful protozoa such as *Giardia cysts* and *Cryptosporidium oocysts* can be almost completely removed under well-controlled conditions. It can also remove about 90 % of bacteria and viruses. Addition of a coagulant before filtration can significantly improve the removal efficiency (Maier et al. 2009). When filtration by ultrafiltration membrane is applied, the removals of bacteria and viruses can achieve 5-log and 2-log, respectively (Wang et al. 2005).

For effluent or reclaimed water disinfection, chlorination and ultraviolet (UV) radiation are common practices. Effective chlorine can bring about damage to bacterial cells mainly by disrupting cell permeability, and to nucleic acids and enzymes targeting the viral capsid, namely the protein coat. The order of resistance to chlorine for different pathogens is usually cysts > viruses > vegetative bacteria by comparing the Ct value (C as disinfectant concentration and t as contact time) corresponding to 99 % (2-log) inactivation. Depending on enteric virus types, there is a very wide variation in their resistance to chlorine (Bitton 2005; Maier et al. 2009).

UV radiation is an alternative disinfection to replace chlorine, for disinfection byproduct control, in many cases. It principally causes thymine and cytosine dimerization to block DNA replication and effectively inactivate target microorganisms. The inactivation effect is proportional to UV dose, but the required UV dose is much different for different microorganisms. In general, the resistance of viruses to UV are considerably higher that bacteria (Bitton 2005; Maier et al. 2009). However, several factors may hinder the application of UV radiation for disinfection, including the higher cost for operation and maintenance, photoreactivation of microorganisms, and the interference of particles (Kollu and Örmeci 2012).

Antibiotic-resistant bacteria are recently drawing wide attention in water environmental studies (Zhang et al. 2009b; LaPara et al. 2011; Rizzo et al. 2013) because they are often found to increase in percentage in the effluent of wastewater treatment comparing with the influent. In order to control the health risk from such kind of bacteria, disinfection may need to be further enhanced (Guo et al. 2013; Huang et al. 2013; Zhang et al. 2014).

3.3.2 *Fecal Indicators Related to Pathogens*

3.3.2.1 Conventional Fecal Indicators and Their Limitations

Human and animals feces are the main source of pathogenic microorganisms in waters. Therefore, fecal indicators have been used for a long time to indicate the presence of enteric pathogens. Conventional fecal indicators include total coliforms, fecal coliforms, *E coli* and fecal streptococci with characteristics as below:

- Total coliforms (TC): Coliform bacteria are a large group of aerobic and facultatively anaerobic, gram-negative, non spore-forming, rod-shaped bacteria, and include *Escherichia*, *Citrobacter*, *Enterobacter*, *Klebsiella* and *Hafnia* species. All members of coliform bacteria can produce gas by lactose fermentation in culture media at 35–37 °C, which is distinguished from other microorganisms. Because of the great number of coliforms in human and warm-blooded animal feces, they are prevalent in sewage with concentration up to 10^7–10^9 CFU/100 mL (Maier et al. 2009). Coliforms are not only found in the intestines of animals but also widely distributed in soil and waters. TC is, therefore, deficient as a fecal indicator because coliforms can grow and multiply in the aquatic environment without fecal pollution. There is also almost no correlation between TC and enteric viruses and protozoa (Zhang et al. 2012b).
- Fecal coliforms and *E. coli*: Temperature is important to the growth of coliforms. A part of coliforms are able to ferment lactose at 44–45 °C and are thus, called fecal coliforms (FC) or thermotolerant coliforms. In most waters, *Escherichia* is the predominant genus of FC, and is called *E. coli* as another fecal indicator, which can be differentiated from other FC by the ability to produce indole from tryptophan or by the production of the enzyme β-glucuronidase. Comparing with TC, FC and *E. coli* are regarded as more suitable indices of fecal pollution. Good relationships are also found between EC or *E. coli* and many pathogenic bacteria such as *Salmonella*, *Shigella* and *Campylobacter*. However, as FC and *E. coli* are detectable in tropical waters without fecal contamination, their suitability to indicate fecal pollution is questioned as well.

- Fecal streptococci (FS): FS is a group of gram-positive, rod-shape bacteria, belonging to the Lancefield group D streptococci. These bacteria are able to grow in 6.5 % sodium chloride at pH 9.6 and 45 °C. According to bacterial taxonomy, fecal streptococci includes some species of two genera, namely the genus *Enterococcus* including *Ent. avium*, *Ent. faecium*, *Ent. durans*, *Ent. faecalis* and *Ent. gallinarium*, and the genus *Streptococcus* including *S. bovis* and *S. equinus* (Maier et al. 2009). *Ent. faecalis* and *Ent. faecium* are more related to human excreta, while *S. bovis* and *S. equines* are prevalent in animals feces. The ratio of fecal coliforms to fecal streptococci (FC/FS ratio) larger or equal to 4 indicates a condition of human feces contamination, and FC/FS ratio below 0.7 indicates a condition of animal pollution (Geldreich and Kenner 1969). However, FC/FS ratio is valid only for 24 h fecal pollution.

3.3.2.2 Alternative Fecal Indicators

Due to the limitations of the conventional fecal indicators as discussed in Sect. 3.3.2.1, the quest for alternative fecal indicators has become an issue of research concern. The suggested alternative fecal indicators include fecal anaerobes and bacteriophage.

- Fecal anaerobes: To overcome the problem of regrowth in aquatic environments when coliform bacteria are used as indicators, fecal anaerobes are recommended as an alternative. In fact, fecal anaerobes are also fecal bacteria but they cannot survive for long in the natural environment due to their low oxygen tolerance. To culture or isolate fecal anaerobes, anaerobic environment has to be provided by using an anaerobic box. The main species of fecal anaerobes as indicator are *Bacteroides* spp., *Bifidobacterium* spp. and *Clostridium perfringens*. Of them, *Bacteroiedes* and *Bifidobacterium* are mostly sourced from feces of warm-blooded animals (Matsuki et al. 2002). By using their characteristics of quick death after being excreted from the host body due to oxygen intolerance, *Bacteroiedes* and *Bifidobacterium* can be good indicators of recent fecal pollution. Moreover, *Bacteriodes* species are highly host-specific (Simpson et al. 2004) and *Bifidobacterium* are primarily associated with humans (Maier et al. 2009). Therefore, they can also be used to discriminate the origin of fecal pollution. In contrast to *Bacteroiedes* and *Bifidobacterium*, although *Clostridium perferingens* species do not multiply in the natural environment as well, they show extreme stability in environmental waters since its hardy spore is resistant to heat and disinfection. As a result, *Clostridium perferingens* is able to indicate past or remote fecal pollution (Savichtcheva and Okabe 2006).
- Bacteriophage: As all known bacterial indicators are unable to well indicate the presence of viruses, bacteriophages are found to be useful as indicators for viruses due to their similarity to enteric viruses and high number in wastewater. Bacteriophages proposed as viral indicators include coliphages and *Bacteroides fragilis* phages. Coliphages use only *E. coli* as hosts for replication and, hence,

can be released into the feces of humans and other warm-blooded animals. There are two major groups of coliphages, namely somatic coliphages, which infect *E. coli* hosts by attaching to the hosts cell wall, and F-RNA coliphages, which initiate infection through fertility (F-) fimbriae of *E. coli* hosts (WHO 2004). Due to their similarity in size, shape structure and genetic constitution to human enteric viruses, F-RNA coliphages are regarded as a more specific index of water quality than somatic coliphage. Comparing with coliphages, *Bacteroides fragilis* bacteriophages are more specific as good indicators for human fecal pollution because of exclusive presence in samples contaminated by human feces. *B. fragilis* bacteriophages are able to survive in surface water but cannot multiply in natural environments. Their decay rate is similar to that of human enteric viruses.

Ideal indicator organism should meet the criteria including similar structure and morphological features to pathogens, high numbers in waters, not growing and multiply in natural environments, not infect humans, and easy to be detected. Unfortunately, the conventional and alternative indicators discussed above are not able to meet all these criteria.

3.3.2.3 Microbial Source Tracking

Alternative indicators, we expect, can also be useful to distinguish the sources of fecal pollution. Therefore, it forms a new field of study—Microbial Source Tracking (MST) or Bacterial Source Tracking (BST), which includes several methodologies used to determine the source of fecal bacteria from samples. Methods for fecal source identification can be divided into culture-based and culture-independent methods. The culture-based methods require growing isolates from environmental samples, while some indicator bacteria such as *Bacteroiedes*, *Bifidobacterium* and *Clostridium perferingens* are difficult to cultivate, which limits their application in great extent. The culture-independent methods allow determining fecal sources by molecular tracers. In these methods, the host-specific 16S ribosomal DNA genetic markers technique is applied well in practice. In this technique, 16S rDNA is extracted from the microorganisms with high host specificity. So, it not only utilizes the characteristics of host-specific microorganisms, but also overcomes their difficult growth in culture media.

16S rDNA markers based on fecal anaerobes were developed to distinguish human and cow fecal pollution, and the *Bacterioides-Prevotella* group was considered as a good indicator (Bernhard and Field 2000). Currently, the main limitation of this approach is the lack of host-specific markers for more species, especially for wildlife (Field and Samadpour 2007). In addition, the characteristics of markers including persistence in waters are poorly understood. Therefore, host-specific markers still need to be further studied before their wide application.

3.3.3 Risk Assessment and Comparison for Water Reuse

3.3.3.1 Waterborne Disease Outbreaks Associated with Water Use

As discussed in Sect. 3.1.2.2, if water reuse for non-potable purposes, such as replenishment of recreational waters and landscape irrigation, is considered, human beings may be under risks of exposure to pathogens through ingestion, inhalation and dermal contact with the reclaimed water. The most reported cases of waterborne diseases outbreaks are related to swimming in recreational waters (Pond 2005; Dwight et al. 2005). Although many outbreaks have occurred in natural waters such as beaches and lakes, there are increasing uses of artificial waters partially or mainly replenished by reclaimed water. The reclaimed water also becomes a potential source of pathogens if not sufficiently disinfected (Asano et al. 2007).

The Centers for Disease Control and Prevention (CDC), USA, reported 789 waterborne disease outbreaks in the period from 1978 to 2010, approximately half of them being gastrointestinal illnesses, while in the period of 2001 to 2010, 235 outbreaks were associated with recreational water (http://www.cdc.gov). Among the outbreaks associated with treated recreational water, 76.2 % were due to *Cryptosporidium* spp. infection because it is extremely chlorine tolerant. Infection by *Shigella* spp. and *E. coli* is often related to untreated recreational water. The reported outbreaks of waterborne diseases due to swimming in recreational waters also include infection with *E. coli* O157:H7 (Keene et al. 1994), *Salmonella* species such *S. jawa* (Levy et al. 1998), *S. enteritidis*, *S. typhimurium* and *S. saint-paul* (Pond 2005), and *Shigella* (Blostein 1991; Keene et al. 1994).

In various waterborne viruses, hepatitis A virus and adenovirus are well reported. There are documented cases of disease outbreaks due to HAV infection related with public pools (Mahoney et al. 1992), and diseases caused by adenovirus in water recreation activities (Caldwell et al. 1974). In most cases, the concentrations of pathogens may not be high, and unevenly distributed in the recreational water. Comparing with potable water use, waterborne disease outbreaks associated with non-potable use may not occur so frequently. However, because of the characteristics of acute infection by pathogens, even an exposure to reclaimed water in a short time may result in a disease outbreak. Therefore, attention should also be paid to the health risk associated with low frequency and low dose of exposure.

3.3.3.2 Outline of Pathogenic Risk Assessment

Human health risks associated with pathogens are usually assessed by two approaches. One is to investigate human health effects caused by pathogens through epidemiologic studies, and another is to evaluate the likelihood of adverse health effects after exposure to pathogenic microorganisms or to a media carrying with pathogens. The so-called pathogenic risk assessment (PRA) refers to the second approach. PRA has been widely applied to the safety control of various waters, including reclaimed water and its reuse.

PRA is principally conducted on the basis of measurement data and mathematical models. There are generally two types of PRA models, namely static model and dynamic model. The static models are mainly used to evaluate the probability of individual infections in single exposure to pathogens in which the number of individuals susceptible to infection is assumed to be time-invariant, and only direct exposure is accounted without consideration of individual immunity to infection. In contrast, more factors are involved in the dynamic models, including categorization of population into various epidemiological states, consideration of time variance of the number of individuals susceptible to infection, and the movement of individuals from one state to another. All these should be taken into account based on epidemiologically relevant data, such as duration of infection and immunity condition because at any time point, only a portion of the population is in a susceptible state, and only those susceptible individuals may be infected through exposure to microorganisms.

Pathogen dose and infectivity are important factors for both static and dynamic models. Therefore, in the identification of pathogens to encounter in water reclamation, assessment of the exposure pathways and quantity, dose-response analysis and characterization, and quantification of the infection or disease are the four basic steps in the procedure of PRA.

3.3.3.3 Exposure Assessment

The target of exposure assessment is to determine the dose of specific pathogens identified to cause diseases, to which humans may be exposed during the process of reclaimed water use. The typical dose per exposure is estimated based on the volume of water ingested, inhaled or contacted with and its pathogen concentration. According to the frequency of exposure to the reclaimed water, the total exposure dose can be evaluated.

Due to various purposes of water reuse and pathways of exposure, the water volume taken in or contacted with is different. Comparing with potable water use, data on exposure during non-potable water use are inadequate, and detailed studies may be required regarding specific cases of reclaimed water use. For reference, Table 3.5 gives the volume of water per exposure for several typical exposure scenarios (Tanaka et al.1998; Metcalf and Eddy 2007).

Table 3.5 Typical exposure scenarios related to reclaimed water

Reuse purpose	Risk group receptor	Volume of water per exposure (mL)	Exposure frequency
Golf course irrigation	Golfer	1	Twice per week
Crop irrigation	Consumer	10	Every day
Recreational impoundment	Swimmer	100	40 days/year (summer season only)
Groundwater recharge	Groundwater consumer	1,000	Every day

Adapted from Tanaka et al. (1998), Metcalf and Eddy (2007)

Although direct or indirect potable use is not the topic of discussion in this book, there are inevitable cases of unconscious ingestion of reclaimed water due to misuse or swimming in recreational waters receiving reclaimed water. These may also need attention. In addition, skin contact of reclaimed water is a pathway of exposure as well. Unfortunately, few reliable data are available by far.

3.3.3.4 Dose-Response Analysis

After understanding the exposure pathways and quantity, dose-response analysis using suitable models is the most important step before quantifying the pathogenic risk. The models widely used for this purpose are the exponential model and beta-Poisson model. The exponential model is derived from an assumption of random occurrence of microorganisms along with a constant probability to initiate infection by a single organism, which expresses the probability of infection, P, as a function of the ingested dose, d, and a coefficient r related to the organism:

$$P = 1 - e^{-rd} \tag{3.3}$$

The beta-Poisson model is an experiential model with two characteristic parameters, in addition to the ingested dose, namely a slope parameter, α, and median infectious dose, N_{50}, and expressed as:

$$P = 1 - \left[1 + \frac{d}{N_{50}}\left(2^{\frac{1}{\alpha}} - 1\right)\right]^{-\alpha} \tag{3.4}$$

As shown in Table 3.6, the exponential model is preferentially used for dose-response analysis for most viruses, while the beta-Poisson model is more suitable for most bacterial pathogens. The recommended model parameters are also listed in Table 3.6.

3.3.3.5 Risk Calculation and Requirement for Safety Control

Risk characterization or quantification is an integration of the information and data obtained from the former steps to draw an overall conclusion on health risk. In most cases, annual risk of infection due to exposure to certain pathogens should be calculated based on a single exposure risk using the following equation:

$$p_a = 1 - (1 - p_i)^n \tag{3.5}$$

where, p_a: annual risk of infection; p_i: single exposure risk; and n: number of exposure events per year.

Table 3.6 Dose-response parameters related to human infection

Organism	Exponential model	Beta-Poisson model		References
	r	α	N_{50}	
Escherichia coli		1.778×10^{-1}	8.60×10^{7}	Haas et al. (1999)
Non-typhoid *Salmonella*		3.126×10^{-1}	2.36×10^{4}	Haas et al. (1999)
Salmonella typhi		1.086×10^{-1}	3.60×10^{6}	Hornick et al. (1966)
Shigella flexnerii and *Shigella dysenteriae*		2.10×10^{-1}	1.12×10^{3}	Haas et al. (1999)
Campylobacter jejuni		1.45×10^{-1}	8.96×10^{2}	Medema et al. (1996)
Vibrio cholerae		2.50×10^{-1}	2.43×10^{2}	Hornick et al. (1971)
Staphylococcus aureus	7.64×10^{-8}			Rose and Haas (1999)
Poliovirus type 1	9.10×10^{-3}			Minor et al. (1981)
Coxsackie virus	1.45×10^{-2}			Couch et al. (1965)
Echovirus type 12	1.28×10^{-2}			Akin (1981)
Rotavirus		2.531×10^{-1}	6.17	Ward et al. (1986)
Adenovirus type 4	6.07×10^{-1}			Couch et al. (1966)
Cryptosporidium	5.72×10^{-2}			Messner et al. (2001)
Giardia lamblia	1.99×10^{-2}			Rose et al. (1991)

When a person ingests a volume of reclaimed water containing a known enteric pathogen concentration, the single exposure dose can be calculated as:

$$d = cV \tag{3.6}$$

where, d: single exposure dose; c: concentration of pathogen in the reclaimed water; and V: volume of water ingested per exposure.

Following the dose-response relationship, p_i is a function of d, and Eq. (3.5) can be written as:

$$p_a = 1 - [1 - f(d)]^n \tag{3.7}$$

where, $f(d)$: dose-response function as Eq. (3.3) or (3.4).

The annual risk of infection, p_a can be constant if the dose, d is invariable. However, in practical cases pathogen concentrations in reclaimed water and the ingested volume may vary, so that p_a may not be a constant, but a variable as the function of dose, d. For a specific case, the distribution of the target pathogens in the reclaimed water can be obtained by long-term water quality monitoring. Using such a distribution and considering the practical condition of exposure, a mathematical function can be developed for expressing the annual risk of infection.

The objective of risk assessment is to set a goal for safety control of reclaimed water, which corresponds to an acceptable risk level. As zero risk is unrealistic, a risk at 10^{-4}/a (one infection per 10,000 persons/year) is thought to be reasonable and acceptable (Haas et al. 1993; Tanaka et al. 1998; Blumenthal et al. 2000). The safety of reclaimed water can thus, be measured as the probability that pathogen infection risk does not exceed the acceptable risk:

$$P(p_a \leq p_a^*) \tag{3.8}$$

where, P: probability of water safety; p_a: pathogen infection risk; p_a^*: acceptable risk.

Combing the above equations, the acceptable single exposure concentration regarding pathogen i can be calculated by Eq. (3.9) or (3.10) using exponential model or beta-Poisson model, respectively.

$$c_i^* = -ln(1 - p_i^*)/rV \tag{3.9}$$

$$c_i^* = N_{50}\left[(1 - p_i^*)^{-\frac{1}{\alpha}} - 1\right]/\left(2^{\frac{1}{\alpha}} - 1\right)V \tag{3.10}$$

where, c_i^*: acceptable single exposure concentration of pathogen i.

Supposing the log removal of pathogen i by treatment is R_i, the concentration of pathogen i in the reclaimed water can be calculated as:

$$\mathrm{c}_i = \mathrm{c}_{i0} 10^{-\mathrm{R}_i} \tag{3.11}$$

where, c_i: concentration of pathogen i in the reclaimed water; c_{i0}: concentration of pathogen i in water to be treated (the secondary effluent if tertiary treatment is taken as the process for reclamation).

As can be seen from the example to be discussed below, c_{i0} in the secondary effluent without disinfection often follows a log-normal distribution with mean μ and standard derivation σ. If c_i also follows the same distribution, then the probability of annual risk of infection below the acceptable risk can be expressed as:

$$\mathrm{P}(\mathrm{p_a} \leq \mathrm{p_a^*}) = \mathrm{P}(\mathrm{c} \leq \mathrm{c_i^*}) = \mathrm{P}(\mathrm{c_{i0}} \leq \mathrm{c_i^*} 10^{\mathrm{R}_i}) = \Phi\left[(\log \mathrm{c_i^*} + \mathrm{R}_i - \mu)/\sigma\right] \tag{3.12}$$

where, Φ: standard normal function.

The required removal efficiency (R_i) of pathogen i for safeguarding water quality can then be determined by the inverse standardized normal function (Φ^{-1}) as:

$$\mathrm{R}_i = \sigma\Phi^{-1}(\mathrm{P}) + \mu - \log \mathrm{c_i^*} \tag{3.13}$$

In order to determine the required removal/inactivation of typical pathogens from secondary effluent in the following stage of physicochemical treatment for water reclamation, a long-term study has been conducted for monitoring selected viruses and pathogenic bacteria in the effluent before disinfection in a domestic wastewater treatment plant in a city in northwestern China, where water reclamation and reuse is required (Zhang and Wang 2014). The target pathogens included infectious enterovirus as typical virus, and *Salmonella typhi*, *Shigella* spp. and *E. coli* as typical bacterial pathogens. Real-time PCR technique was applied for

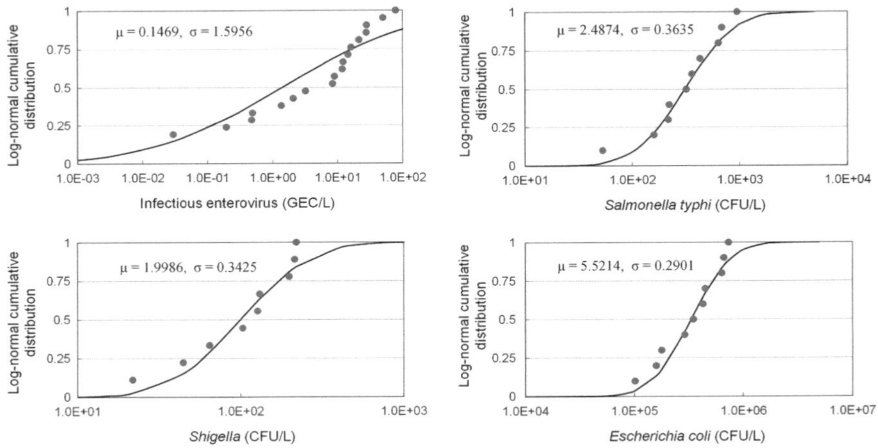

Fig. 3.9 Log-normal cumulative distribution of pathogens in the secondary effluent. Reprinted by permission from Taylor & Francis: Zhang and Wang (2014), copyright 2014

pathogen detection with universal primers for enterovirus and specific primers for each of the bacterial pathogens. As shown in Fig. 3.9, the concentration of each pathogen was found to follow the log-normal relation well, so that by data manipulation log-normal distribution curves could be obtained with their characteristic parameters as the mean, μ and standard derivation, σ shown in the figure.

Following the methods discussed in Sect. 3.3.3, human health risk assessment was conducted regarding two typical purposes of reclaimed water use in an urban area, namely golf course irrigation and recreational impoundment. Tables 3.5 and 3.6 were referred for determining the exposure volume and frequency, and dose-response models, and their parameters under the assumptions that enteroviruses were polioviruses, coxsakieviruses, or echoviruses because enterovirus is a virus group and there is no definite dose-response relationship for it. The acceptable risk for each pathogen was set as 10^{-4}/a, and the output of risk assessment was expressed as the reliability of risk control to the acceptable level.

Figure 3.10 shows the relation between log removal of various pathogens and the reliability of reclaimed water safety control. Regarding enteroviruses, regardless of whether they are assumed to be polioviruses, coxsakieviruses or echoviruses, the curves shown in Fig. 3.10 are not very different from each other, indicating that the dose-response relationships of the three viruses can all be used for risk assessment for the enterovirus group. With increasing log removal, the reliability of safety control gradually increases, and the removals for completely reliable control, namely reliability near 100 %, have to be higher than 5-log and 7-log for golf course irrigation and recreational impoundment, respectively. If 95 % is taken as a reasonable reliability, then about 4-log and 5.5-log removals would be required, respectively, for the two reuse purposes.

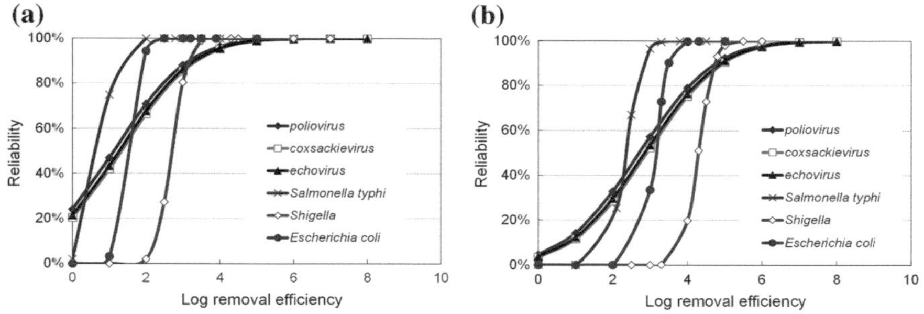

Fig. 3.10 Relations between log removal of pathogens and the reliability of reclaimed water safety control regarding two typical reuse purposes of golf course irrigation (**a**) and recreational impoundment (**b**). Reprinted by permission from Taylor & Francis: Zhang and Wang (2014), copyright 2014

Comparing with viruses, the log removal–reliability curves vary more sharply, namely a quick rise in the reliability with increasing log removal in certain ranges, but the three bacteria show noticeable differences in the requirement for log removal to achieve a given reliability. The log removal for *Shigella* is the highest for both reuse purposes and then followed by *E. coli*. If 95 % is also taken as the reasonable reliability for controlling bacterial pathogens, the required removals would be 3.3-log, 2-log, and 1.4-log, respectively for *Shigella*, *E. coli*, and *Salmonella typhi*, in the case of golf course irrigation, while it would be 4.9-log, 3.6-log, and 2.9-log, respectively, for the three bacteria in the case of recreational impoundment.

References

Akin EW (1981) Paper presented at the US EPA symposium on microbial health considerations of soil disposal of domestic wastewaters

Allen LM, Briggle TV, Pfaffenberger CD (1982) Absorption and excretion of cyanuric acid in long-distance swimmers. Drug Metab 13(3):499–516

Alturki AA, Tadkaew N, McDonald JA, Khan SJ, Price WE, Nghiem LD (2010) Combining MBR and NF/RO membrane filtration for the removal of trace organics in indirect potable water reuse applications. J Membr Sci 365:206–215. doi:10.1016/j.memsci.2010.09.008

Asano T, Burton FL, Leverenz HL, Tsuchihashi R, Tchobanoglous G (2007) Water reuse: issues, technologies, and applications. Metcalf & Eddy, McGraw-Hill, New York

AWWA (2006) Waterborne pathogens, 2nd edn. American Water Works Association, Denver

Backhaus T, Froehner K, Altenburger R, Grimme LH (1997) Toxicity testing with Vibrio Fischeri: a comparison between the long term (24 h) and the short term (30 min) bioassay. Chemosphere 35(12):2925–2938. doi:10.1016/S0045-6535(97)00340-8

Beker VWA, Wolterbeek A, te Brake L, Snel C, Menke A, Rubingh C, de Groot D, Kroese D (2013) A category approach to predicting the developmental (neuro) toxicity of organotin compounds: the value of the zebrafish (Danio rerio) embryotoxicity test (ZET). Reprod Toxicol 41:35–44. doi:10.1016/j.reprotox.2013.06.067

Beker VWA, Snel C, Rijkmans E, de Groot D, Bouma M, Hermsen S, Piersma A, Menke A, Wolterbeek A (2014) Zebrafish embryotoxicity test for developmental (neuro) toxicity: demo case of an integrated screening approach system using anti-epileptic drugs. Reprod Toxicol 49:101–116. doi:10.1016/j.reprotox.2014.07.082

Bernhard AE, Field KG (2000) A PCR assay to discriminate human and ruminant feces on the basis of host differences in *Bacteroides Prevotella* genes encoding 16S rRNA. Appl Environ Microbiol 66(10):4571–4574. doi:10.1128/AEM.66.10.4571-4574.2000

Bitton G (2005) Wastewater microbiology. Wiley, New York

Blankenship AL, Kay DP, Zwiernik MJ, Holem RR, Newsted JL, Hecker M, Giesy JP (2008) Toxicity reference values for mink exposed to 2,3,7,8-tetrachlodibenzo-p-dioxin (TCDD) equivalents (TEQs). Ecotoxicol Environ Saf 69(3):325–349. doi:10.1016/j.ecoenv.2007.08.017

Blostein J (1991) Shigellosis from swimming in a park pond in Michigan. Public Health Rep 106:317–322

Blumenthal UJ, Mara DD, Peasey A, Ruiz PG, Stott R (2000) Guidelines for the microbiological quality of treated wastewater used in agriculture: recommendations for revising WHO guidelines. Bull WHO 78(9):1104–1116

Boyce TG, Swerdlow DL, Griffin PM (1995) *Escherichi coli* O157:H7 and the hemolytic uremic syndrome. N Engl J Med 333:364–368. doi:10.1056/NEJM199508103330608

Cabeza Y, Candela L, Ronen D, Teijon G (2012) Monitoring the occurrence of emerging contaminants in treated wastewater and groundwater between 2008 and 2010. The Baix Llobregat (Barcelona, Spain). J Hazard Mater 239–240(0):32–39. doi:10.1016/j.jhazmat.2012.07.032

Caldwell GG, Lindsey NJ, Wulff H, Donnelly DD, Bohl FN (1974) Epidemic with adenovirus type 7 acute conjuctivitis in swimmers. Am J Epidemiol 99:230–234

Camann DE, Moore BE (1988) Viral infections based on clinical sampling at a spray irrigation site. Implementing water reuse. In: Proceedings of water reuse symposium IV. AWWA Research Foundation, Denver, pp 847–863

Cavaco SA, Fernandes S, Quina MM, Ferreira LM (2007) Removal of chromium from electroplating industry effluents by ion exchange resins. J Hazard Mater 144(3):634–638. doi:10.1016/j.jhazmat.2007.01.087

Chen W, Lu S, Peng C, Jiao W, Wang M (2013) Accumulation of Cd in agricultural soil under long-term reclaimed water irrigation. Environ Pollut 178:294–299. doi:10.1016/j.envpol.2013.03.036

Chen W, Lu S, Pan N, Wang Y, Wu L (2015) Impact of reclaimed water irrigation on soil health in urban green areas. Chemosphere 119:654–661. doi:10.1016/j.chemosphere.2014.07.035

Cliver DO (1985) Vehicular transmission of hepatitis. Annu Rev Publ Health 13:235–292

Couch RB, Cate T, Gerone P, Fleet W, Lang D, Griffith WR, Knight V (1965) Production of illness with a small-particle aerosol of Coxsackie A21. J Clin Invest 44(4):535–542. doi:10.1172/JCI105166

Couch RB, Cate T, Gerone P, Fleet W, Lang D, Griffith WR, Knight V (1966) Production of illness with a small-particle aerosol of Adenovirus type 4. Bact Rev 30:517–528

Craun GF (2012) The importance of waterborne disease outbreak surveillance in the United States. Ann Ist Super Santà 48(4):447–459. doi:10.4415/ANN_12_04_11

Craun GF, Brunkard JM, Yoder JS, Roberts VA, Carpenter J, Wade T, Calderon RL, Roberts JM, Beach MJ, Roy SL (2010) Causes of outbreaks associated with drinking water in the United States from 1971 to 2006. Clin Microb Rev 2010 23(3):507–528. doi:10.1128/CMR.00077-09

Crump JA, Luby SP, Mintz ED (2004) The global burden of typhoid fever. Bull World Health Organ 82:346–353. doi:10.1590/S0042-96862004000500008

De Koning J, Bixio D, Karabelas A, Salgot M, Schäfer A (2008) Characterisation and assessment of water treatment technologies for reuse. Desalination 218(1):92–104. doi:10.1016/j.desal.2006.08.024

Deniz F, Sadhwani JJ, Veza JM (2010) New quality criteria in wastewater reuse—the case of Gran Canaria. Desalination 250:716–722. doi:10.1016/j.desal.2008.11.029

Dodder NG, Maruya KA, Lee Ferguson P, Grace R, Klosterhaus S, La Guardia MJ, Lauenstein GG, Ramirez J (2014) Occurrence of contaminants of emerging concern in mussels (Mytilus spp.) along the California coast and the influence of land use, storm water discharge, and treated wastewater effluent. Mar Pollut Bull 81(2):340–346. doi:10.1016/j.marpolbul.2013.06.041

Dufour AP, Evans O, Behymer TD, Cantu R (2006) Water ingestion during swimming activities in a pool: a pilot study. J Water Health 4(4):425–430

Dunlap PV (2009) Bioluminescence, microbial. Encyclopedia of microbiology, 3rd edn. Academic Press, Oxford, pp 45–61

Dwight RH, Fernandez LM, Baker DB, Semenza JC, Olson BH (2005) Estimating the economic burden from illnesses associated with recreational water pollution—a case study in Orange County, California. J Environ Manage 76(2):95–103. doi:10.1016/j.jenvman.2004.11.017

Ellis JB (2006) Pharmaceutical and personal care products (PPCPs) in urban receiving waters. Environ Pollut 144:184–189. doi:10.1016/j.envpol.2005.12.018

Estévez E, Cabrera MdC, Molina DA, Robles MJ, Palacios DM (2012) Screening of emerging contaminants and priority substances (2008/105/EC) in reclaimed water for irrigation and groundwater in a volcanic aquifer (Gran Canaria, Canary Islands, Spain). Sci Total Environ 433(0):538–546. doi:10.1016/j.scitotenv.2012.06.031

Farré M, Barceló D (2003) Toxicity testing of wastewater and sewage sludge by biosensors, bioassays and chemical analysis. TrAC Trend Anal Chem 22(5):299–310. doi:10.1016/s0165-9936(03)00504-1

Field KG, Samadpour M (2007) Fecal source tracking, the indicator paradigm, and managing water quality. Water Res 41(16):3517–3538. doi:10.1016/j.watres.2007.06.056

Fleisher LA (2012) Anesthesia and uncommon diseases, 6th edn. Saunders Elsevier, Philadelphia

Froehner K, Meyer W, Grimme LH (2002) Time-dependent toxicity in the long-term inhibition assay with Vibrio fischeri. Chemosphere 46(7):987–997. doi:10.1016/s0045-6535(01)00209-0

Fujioka T, Khan S, Poussade Y, Drewes JE, Nghiem LD (2012) N-nitrosamine removal by reverse osmosis for indirect potable water reuse—a critical review based on observations from laboratory-, pilot- and full-scale studies. Sep Purif Technol 98:503–515. doi:10.1016/j.seppur.2012.07.025

Fulladosa E, Murat JC, Martínez M, Villaescusa I (2005) Patterns of metals and arsenic poisoning in vibrio fischeri bacteria. Chemosphere 60(1):43–48. doi:10.1016/j.chemosphere.2004.12.026

Gao Y, Lin ZF, Chen R, Wang T, Liu SS, Yao ZF, Yin DQ (2012) Using molecular docking to compare toxicity of reactive chemicals to freshwater and marine luminous bacteria. Mol Inform 31(11–12):809–816. doi:10.1002/minf.201200060

Ge HL, Liu SS, Zhu XW, Liu HL, Wang LJ (2011) Predicting hormetic effects of ionic liquid mixtures on luciferase activity using the concentration addition model. Environ Sci Technol 45 (4):1623–1629. doi:10.1021/es1018948

Geldreich EE, Kenner BA (1969) Comments on fecal streptococci in stream pollution. J Water Pollut Control Fed 41:R336–R341

Girotti S, Ferri EN, Fumo MG, Maiolini E (2008) Monitoring of environmental pollutants by bioluminescent bacteria. Anal Chim Acta 608(1):2–29. doi:10.1016/j.aca.2007.12.008

Goodbred SL, Bryant WL, Rosen MR, Alvarez D, Spencer T (2009) How useful are the "other" semipermeable membrane devices (SPMDs); the mini-unit (15.2 cm long)? Sci Total Environ 407(13):4149–4156. doi:10.1016/j.scitotenv.2009.02.037

Guilhermino L, Diamantino T, Carolina SM, Soares AMVM (2000) Acute toxicity test with daphnia magna: an alternative to mammals in the prescreening of chemical toxicity? Ecotoxicol Environ Saf 46(3):357–362. doi:10.1006/eesa.2000.1916

Guo MT, Yuan QB, Yang J (2013) Microbial selectivity of UV treatment on antibiotic-resistant heterotrophic bacteria in secondary effluents of a municipal wastewater treatment plant. Water Res 47(16):6388–6394. doi:10.1016/j.watres.2013.08.012

Haas CN, Rose JB, Gerba CP, Regli S (1993) Risk assessment of viruses in drinking water. Risk Anal 13:545–552. doi:10.1111/j.1539-6924.1993.tb00013.x

Haas CN, Rose JB, Gerba CP (1999) Quantitative microbial risk assessment. Wiley, New York

Hansen PD (2007) Risk assessment of emerging contaminants in aquatic systems. TrAC Trend Anal Chem 26(11):1095–1099. doi:10.1016/j.trac.2007.10.001
Harguinteguy CA, Cirelli AF, Pignata ML (2014) Heavy metal accumulation in leaves of aquatic plant Stuckenia filiformis and its relationship with sediment and water in the Suquía river (Argentina). Microchem J 114:111–118. doi:10.1016/j.microc.2013.12.010
Hegedüs A, Erdei S, Horváth G (2001) Comparative studies of H_2O_2 detoxifying enzymes in green and greening barley seedlings under cadmium stress. Plant Sci 160(6):1085–1093. doi:10.1016/S0168-9452(01)00330-2
Hernando MD, Fernández AAR, Tauler R, Barceló D (2005) Toxicity assays applied to wastewater treatment. Talanta 65(2):358–366. doi:10.1016/j.talanta.2004.07.012
Herrero O, Pérez MJM, Fernández FP, Carvajal LL, Peropadre A, Hazen MJ (2012) Toxicological evaluation of three contaminants of emerging concern by use of the Allium cepa test. Mutat Res/Gen Tox Env 743(1–2):20–24. doi:10.1016/j.mrgentox.2011.12.028
Hirata K, Tsuji N, Miyamoto K (2005) Biosynthetic regulation of phytochelatins, heavy metal-binding peptides. J Biosci Bioeng 100(6):593–599. doi:10.1263/jbb.100.593
Hong B, Garabrant D, Hedgeman E, Demond A, Gillespie B, Chen Q, Chang CW, Towey T, Knutson K, Franzblau A, Lepkowski J, Adriaens P (2009) Impact of WHO 2005 revised toxic equivalency factors for dioxins on the TEQs in serum, household dust and soil. Chemosphere 76(6):727–733. doi:10.1016/j.chemosphere.2009.05.034
Hornick RB, Woodward TE, McCrumb FR, Dawkin AT, Snyder MJ, Bulkeley JT, Macorra FDL, Corozza FA (1966) Study of induced typhoid fever in man. I. Evaluation of vaccine effectiveness. Tran Assoc Am Phys 79:361–367
Hornick RB, Music SI, Wenzel R, Cash RA, Libonati JP, Snyder MJ, Woodward TE (1971) The broad street pump revisited: response of volunteers to ingested cholera vibrios. Bull NY Acad Med 47(10):1181–1191
Huang JJ, Hu HY, Wu YH, Wei B, Lu Y (2013) Effect of chlorination and ultraviolet disinfection on tet A-mediated tetracycline resistance of Escherichia coli. Chemosphere 90(8):2247–2253. doi:10.1016/j.chemosphere.2012.10.008
Hussain MM, Amanchi NR, Solanki VR, Bhagavathi M (2008) Low cost microbioassay test for assessing cytopathological and physiological responses of ciliate model Paramecium caudatum to carbofuran pesticide. Pestic Biochem Phys 90(1):66–70. doi:10.1016/j.pestbp.2007.07.006
Ince NH, Apikyan IG (2000) Combination of activated carbon adsorption with light-enhanced chemical oxidation via hydrogen peroxide. Water Res 34(17):4169–4176. doi:10.1016/S0043-1354(00)00194-9
Jenkins RO, Craig PJ, Francesconi KA, Harrington CF (2007) 12.13-Environmental and biological aspects of organometallic compounds. Comprehensive organometallic chemistry III. Elsevier, Oxford, pp 603–661
Kadokami K, Li X, Pan S, Ueda N, Hamada K, Jinya D, Iwamura T (2013) Screening analysis of hundreds of sediment pollutants and evaluation of their effects on benthic organisms in Dokai Bay, Japan. Chemosphere 90(2):721–728. doi:10.1016/j.chemosphere.2012.09.055
Kasprzyk-Hordern B, Dinsdale RM, Guwy AJ (2008) The occurrence of pharmaceuticals, personal care products, endocrine disruptors and illicit drugs in surface water in South Wales, UK. Water Res 42(13):3498–3518. doi:10.1016/j.watres.2008.04.026
Kayikcioglu HH (2012) Short-term effects of irrigation with treated domestic wastewater on microbiological activity of a vertic xerofluvent soil under Mediterranean conditions. J Environ Manage 102:108–114. doi:10.1016/j.jenvman.2011.12.034
Keene WE, McAnulty JM, Hoesly FC, Williams LP Jr, Hedberg K, Oxman GL, Barrett TJ, Pfaller MA, Fleming DW (1994) A swimming associated outbreak of hemorrhagic colitis caused by Escherichia coli O157:H7 and Shigella sonnei. New Engl J Med 331:579–584. doi:10.1056/NEJM199409013310904
Kokkali V, van Delft W (2014) Overview of commercially available bioassays for assessing chemical toxicity in aqueous samples. TrAC Trend Anal Chem 61:133–155. doi:10.1016/j.trac.2014.08.001

Kollu K, Örmeci B (2012) Effect of particles and bioflocculation on ultraviolet disinfection of Escherichia coli. Water Res 46(3):750–760. doi:10.1016/j.watres.2011.11.046

Komaitis E, Vasiliou E, Kremmydas G, Georgakopoulos DG, Georgiou C (2010) Development of a fully automated flow injection analyzer implementing bioluminescent biosensors for water toxicity assessment. Sensors 10(8):7089–7098. doi:10.3390/s100807089

LaPara TM, Burch TR, McNamara PJ, Tan DT, Yan M, Eichmiller JJ (2011) Tertiary-treated municipal wastewater is a significant point source of antibiotic resistance genes into duluth-superior harbor. Environ Sci Technol 45(22):9543–9549. doi:10.1021/es202775r

Ledakowicz S, Solecka M, Zylla R (2001) Biodegradation, decolourisation and detoxification of textile wastewater enhanced by advanced oxidation processes. J Biotechnol 89(2):175–184. doi:10.1016/S0168-1656(01)00296-6

Lee WM, Ha SW, Yang CY, Lee JK, An YJ (2011) Effect of fluorescent silica nanoparticles in embryo and larva of oryzias latipes: sonic effect in nanoparticle dispersion. Chemosphere 82 (3):451–459. doi:10.1016/j.chemosphere.2010.09.055

Leusch FDL, Khan SJ, Gagnon MM, Quayle P, Trinh T, Coleman H, Rawson C, Chapman HF, Blair P, Nice H, Reitsema T (2014a) Assessment of wastewater and recycled water quality: a comparison of lines of evidence from in vitro, in vivo and chemical analyses. Water Res 50:420–431. doi:10.1016/j.watres.2013.10.056

Leusch FDL, Khan SJ, Laingam S, Prochazka E, Froscio S, Trinh T, Chapman HF, Humpage A (2014b) Assessment of the application of bioanalytical tools as surrogate measure of chemical contaminants in recycled water. Water Res 49:300–315. doi:10.1016/j.watres.2013.11.030

Levantesi C, Bonadonna L, Briancesco R, Grohmann E, Toze S, Tandoi V (2012) Salmonella in surface and drinking water: occurrence and water-mediated transmission. Food Res Int 45 (2):587–602. doi:10.1016/j.foodres.2011.06.037

Levy DA, Bens MS, Craun GF, Calderon RL, Herwaldt BL (1998) Surveillance for waterborne-disease outbreaks—United States, 1995–1996. MMWR 47:1–34

Liu SS, Song XQ, Liu HL, Zhang YH, Zhang J (2009a) Combined photobacterium toxicity of herbicide mixtures containing one insecticide. Chemosphere 75:381–388. doi:10.1016/j.chemosphere.2008.12.026

Liu YJ, Zhang CM, Wang XC (2009b) Simultaneous detection of enteric bacteria from surface waters by QPCR in comparison with conventional bacterial indicators. Environ Monit Assess 158(1–4):535–544. doi:10.1007/s10661-008-0602-1

Lopez RR, Kazlauskaite L, Ribo J, Riva MC, Gonzalez S, Cortina JL (2012) Evaluation of an automated luminescent bacteria assay for in situ aquatic toxicity determination. Sci Total Environ 440:307–313. doi:10.1016/j.scitotenv.2012.05.043

Łuczkiewicz A, Jankowska K, Fudala KS, Olańczuk NK (2010) Antimicrobial resistance of fecal indicators in municipal wastewater treatment plant. Water Res 44(17):5089–5097. doi:10.1016/j.watres.2010.08.007

Luo L, Ke C, Guo X, Shi B, Huang M (2014) Metal accumulation and differentially expressed proteins in gill of oyster (Crassostrea hongkongensis) exposed to long-term heavy metal-contaminated estuary. Fish Shellfish Immun 38(2):318–329. doi:10.1016/j.fsi.2014.03.029

Ma XY, Wang XC (2013) Ecotoxicity comparison of organic contaminants and heavy metals using Vibrio-qinghaiensis sp.-Q67. Water Sci Technol 67(10):2221–2227. doi:10.2166/wst.2013.113

Ma XY, Wang XC, Liu YJ (2011) Study of the variation of ecotoxicity at different stages of domestic wastewater treatment using Vibrio-qinghaiensis sp.-Q67. J Hazard Mater 190(1–3):100–105. doi:10.1016/j.jhazmat.2011.03.003

Ma XY, Wang XC, Liu YY (2012) Cytotoxicity and genotoxicity evaluation of urban surface waters using freshwater luminescent bacteria Vibrio-qinghaiensis sp.-Q67 and Vicia faba root tip. J Environ Sci 24(10):1861–1866. doi:10.1016/S1001-0742(11)61012-7

Ma XY, Wang XC, Ngo HH, Guo W, Wu MN, Wang N (2013) Reverse osmosis pretreatment method for toxicity assessment of domestic wastewater using Vibrio qinghaiensis sp.-Q67. Ecotoxicol Environ Saf 97(0):248–254. doi:10.1016/j.ecoenv.2013.08.001

Ma XY, Wang XC, Ngo HH, Guo W, Wu MN, Wang N (2014) Bioassay based luminescent bacteria: interferences, improvements, and applications. Sci Total Environ 468–469:1–11. doi:10.1016/j.scitotenv.2013.08.028

Macova M, Escher BI, Reungoat J, Carswell S, Chue KL, Keller J, Mueller JF (2010) Monitoring the biological activity of micropollutants during advanced wastewater treatment with ozonation and activated carbon filtration. Water Res 44(2):477–492. doi:10.1016/j.watres.2009.09.025

Mahmoud IY, Al-Bahry SN, Al-Musharafi SK (2013) Fresh water habitat pollution by treated sewage effluent in relation to multiple-antibiotic-resistant bacteria. APCBEE Procedia 5:363–367. doi:10.1016/j.apcbee.2013.05.062

Mahoney FJ, Farley TA, Kelso KY, Wilson SA, Horan JM, McFarland LM (1992) An outbreak of hepatitis a associated with swimming in a public pool. J Infect Dis 165:613–618

Maier RM, Pepperl L, Gerba CP (2009) Environmental microbiology, 2nd edn. Elsevier Science, San Diego

Matsuki T, Watanabe K, Fujimoto J, Miyamoto Y, Takada T, Matsumoto K, Oyaizu H, Tanaka R (2002) Development of 16S rRNA-gene-targeted group-specific primers for analysis of human intestinal bifidobacteria. Appl Environ Microbiol 68:5445–5451. doi:10.1128/AEM.68.11.5445-5451.2002

Maunula L, Miettinen IT, von Bonsdorff CH (2005) Norovirus outbreaks from drinking water. Emerg Infect Dis 11(11):1716–1721. doi:10.3201/eid1111.050487

Medema GJ, Teunis PFM, Havelaar AH, Haas CN (1996) Assessment of the dose-response relationship of Campylobacter jejuni. Int J Food Microbiol 39:101–112. doi:10.1016/0168-1605(96)00994-4

Messner MJ, Chappell CL, Okhuysen PC (2001) Risk assessment for cryptosporidium: a hierarchical Bayesian analysis of human dose response data. Water Res 35(16):3934–3940. doi:10.1016/S0043-1354(01)00119-1

Metcalf, Eddy (2007) Water reuse: issues, technologies and applications. McGraw-Hill Publisher, New York

Minor TE, Allen CI, Tsiatis AA, Nelson DD, D'Alessio DJ (1981) Human infective dose determination for oral poliovirus type 1 vaccine in infants. J Clin Microbiol 13:388

Ml Farré, García MJ, Tirapu L, Ginebreda A, Barceló D (2001) Wastewater toxicity screening of non-ionic surfactants by Toxalert® and Microtox® bioluminescence inhibition assays. Anal Chim Acta 427(2):181–189. doi:10.1016/s0003-2670(00)01022-9

Mohanpuria P, Rana NK, Yadav SK (2007) Cadmium induced oxidative stress influence on glutathione metabolic genes of Camellia sinensis (L.) O. Kuntze. Environ Toxicol 22(4):368–374. doi:10.1002/tox.20273

Monteiro L, Brinke M, dos Santos G, Traunspurger W, Moens T (2014) Effects of heavy metals on free-living nematodes: a multifaceted approach using growth, reproduction and behavioural assays. Eur J Soil Biol 62:1–7. doi:10.1016/j.ejsobi.2014.02.005

Mori IC, Arias BCR, Koutsaftis A, Ogo A, Kawano T, Yoshizuka K, Inayat HSH, Aoyama I (2015) Toxicity of tetramethylammonium hydroxide to aquatic organisms and its synergistic action with potassium iodide. Chemosphere 120:299–304. doi:10.1016/j.chemosphere.2014.07.011

Morugán-Coronado A, García OF, Mataix SJ, Arcenegui V, Mataix BJ (2011) Short-term effects of treated wastewater irrigation on Mediterranean calcareous soil. Soil Till Res 112(1):18–26. doi:10.1016/j.still.2010.11.004

Musolff A, Leschik S, Möder M, Strauch G, Reinstorf F, Schirmer M (2009) Temporal and spatial patterns of micropollutants in urban receiving waters. Environ Pollut 157(11):3069–3077. doi:10.1016/j.envpol.2009.05.037

NAS (1983) Risk assessment in federal government: managing the process. National Academy Press, Washington

Omoike AI, Vanloon GW (1999) Removal of phosphorus and organic matter removal by alum during wastewater treatment. Water Res 33(17):3617–3627. doi:10.1016/S0043-1354(99)00075-5

OW/ORD Emerging Contaminants Workgroup (2008). White paper aquatic life criteria for contaminants of emerging concern. Emerging contaminants workgroup, Part 1 general challenges and recommendation

Palacios-Díaz MP, Mendoza GV, Fernández VJR, Rodríguez RF, Tejedor JMT, Hernández MJM (2009) Subsurface drip irrigation and reclaimed water quality effects on phosphorus and salinity distribution and forage production. Agric Water Manage 96(11):1659–1666. doi:10.1016/j.agwat.2009.06.021

Parvez S, Venkataraman C, Mukherji S (2006) A review on advantages of implementing luminescence inhibition test (Vibrio fischeri) for acute toxicity prediction of chemicals. Environ Int 32(2):265–268. doi:10.1016/j.envint.2005.08.022

Pavlaki MD, Pereira R, Loureiro S, Soares AMVM (2011) Effects of binary mixtures on the life traits of daphnia magna. Ecotoxicol Environ Saf 74(1):99–110. doi:10.1016/j.ecoenv.2010.07.010

Pond K (2005) Water recreation and disease. Plausibility of associated Infections: acute effects, sequelae and mortality. IWA publishing on behalf of the World Health Organization, London

Prat N, Rieradevall M, Barata C, Munné A (2013) The combined use of metrics of biological quality and biomarkers to detect the effects of reclaimed water on macroinvertebrate assemblages in the lower part of a polluted Mediterranean river (Llobregat River, NE Spain). Ecol Indic 24:167–176. doi:10.1016/j.ecolind.2012.06.010

Qin LT, Liu SS, Zhang J, Xiao QF (2011) A novel model integrated concentration addition with independent action for the prediction of toxicity of multi-component mixture. Toxicol 280(3):164–172. doi:10.1016/j.tox.2010.12.007

Rahman MF, Yanful EK, Jasim SY (2009a) Endocrine disrupting compounds (EDCs) and pharmaceuticals and personal care products (PPCPs) in the aquatic environment: implications for the drinking water industry and global environmental health. J Water Health 7(2):224–243. doi:10.2166/wh.2009.021

Rahman MF, Yanful EK, Jasim SY (2009b) Occurrences of endocrine disrupting compounds and pharmaceuticals in the aquatic environment and their removal from drinking water: challenges in the context of the developing world. Desalination 248(1–3):578–585. doi:10.1016/j.desal.2008.05.105

Ratola N, Cincinelli A, Alves A, Katsoyiannis A (2012) Occurrence of organic microcontaminants in the wastewater treatment process. A mini review. J Hazard Mater 239–240:1–18. doi:10.1016/j.jhazmat.2012.05.040

Reinthaler FF, Posch J, Feierl G, Wüst G, Haas D, Ruckenbauer G, Mascher F, Marth E (2003) Antibiotic resistance of E. coli in sewage and sludge. Water Res 37(8):1685–1690. doi:10.1016/S0043-1354(02)00569-9

Rico D, Martín GA, Díaz S, de Lucas P, Gutiérrez JC (2009) Heavy metals generate reactive oxygen species in terrestrial and aquatic ciliated protozoa. Comp Biochem Phys C 149(1):90–96. doi:10.1016/j.cbpc.2008.07.016

Rizzo L (2011) Bioassays as a tool for evaluating advanced oxidation processes in water and wastewater treatment. Water Res 45(15):4311–4340. doi:10.1016/j.watres.2011.05.035

Rizzo L, Manaia C, Merlin C, Schwartz T, Dagot C, Ploy MC, Michael I, Fatta KD (2013) Urban wastewater treatment plants as hotspots for antibiotic resistant bacteria and genes spread into the environment: a review. Sci Total Environ 447:345–360. doi:10.1016/j.scitotenv.2013.01.032

Rodriguez C, Linge K, Blair P, Busetti F, Devine B, Van Buynder P, Weinstein P, Cook A (2012) Recycled water: potential health risks from volatile organic compounds and use of 1,4-dichlorobenzene as treatment performance indicator. Water Res 46(1):93–106. doi:10.1016/j.watres.2011.10.032

Roig N, Nadal M, Sierra J, Ginebreda A, Schuhmacher M, Domingo JL (2011) Novel approach for assessing heavy metal pollution and ecotoxicological status of rivers by means of passive sampling methods. Environ Int 37(4):671–677. doi:10.1016/j.envint.2011.01.007

Rose JB, Haas CN (1999) A risk assessment framework for the evaluation of skin infections and the potential impact of antibacterial soap washing. Am J Infect Control 27(6):S26–S33. doi:10.1016/S0196-6553(99)70039-8

Rose JB, Haa CN, Regli S (1991) Risk assessment and the control of waterborne giardiasis. Am J Public Health 81:709–713. doi:10.2105/AJPH.81.6.709

Santasmasas C, Rovira M, Clarens F, Valderrama C (2013) Grey water reclamation by decentralized MBR prototype. Resour Conserv Recy 72:102–107. doi:10.1016/j.resconrec.2013.01.004

Savichtcheva O, Okabe S (2006) Alternative indicators of fecal pollution: relations with pathogens and conventional indicators, current methodologies for direct pathogen monitoring and future application perspectives. Water Res 40(13):2463–2476. doi:10.1016/j.watres.2006.04.040

Sharma BM, Bharat GK, Tayal S, Nizzetto L, Čupr P, Larssen T (2014) Environment and human exposure to persistent organic pollutants (POPs) in India: a systematic review of recent and historical data. Environ Int 66:48–64. doi:10.1016/j.envint.2014.01.022

Shaw M, Negri A, Fabricius K, Mueller JF (2009) Predicting water toxicity: pairing passive sampling with bioassays on the Great Barrier Reef. Aquat Toxicol 95(2):108–116. doi:10.1016/j.aquatox.2009.08.007

Silva A, Figueiredo SA, Sales MG, Delerue MC (2009) Ecotoxicity tests using the green algae Chlorella vulgaris—a useful tool in hazardous effluents management. J Hazard Mater 167(1–3):179–185. doi:10.1016/j.jhazmat.2008.12.102

Simpson JM, Santo Domingo JW, Reasoner DJ (2004) Assessment of equine fecal contamination: the search for alternative bacterial source-tracking targets. FEMS Microbiol Ecol 47(1):65–75. doi:10.1016/S0168-6496(03)00250-2

Smital T, Terzic S, Zaja R, Senta I, Pivcevic B, Popovic M, Mikac I, Tollefsen KE, Thomas KV, Ahel M (2011) Assessment of toxicological profiles of the municipal wastewater effluents using chemical analyses and bioassays. Ecotoxicol Environ Saf 74(4):844–851. doi:10.1016/j.ecoenv.2010.11.010

Smith T (2011) Overcoming challenges in wastewater reuse: a case study of San Antonio, Texas. Harvard College Cambridge, Mass

Sonune A, Ghate R (2004) Development in wastewater treatment methods. Desalination 167:55–63. doi:10.1016/j.desal.2004.06.113

Soupilas A, Papadimitriou CA, Samaras P, Gudulas K, Petridis D (2008) Monitoring of industrial effluent ecotoxicity in the greater Thessaloniki area. Desalination 224(1–3):261–270. doi:10.1016/j.desal.2007.07.003

Suhendrayatna, Ohki A, Nakajima T, Maeda S (2002) Studies on the accumulation and transformation of arsenic in freshwater organisms I. Accumulation, transformation and toxicity of arsenic compounds on the Japanese Medaka, Oryzias latipes. Chemosphere 46(2):319–324. doi:10.1016/S0045-6535(01)00084-4

Sun Y, Huang H, Sun Y, Wang C, Shi XL, Hu HY, Kameya T, Fujie K (2013) Ecological risk of estrogenic endocrine disrupting chemicals in sewage plant effluent and reclaimed water. Environ Pollut 180:339–344. doi:10.1016/j.envpol.2013.05.006

Takahashi T, Yoshii M, Kawano T, Kosaka T, Hosoya H (2005) A new approach for the assessment of acrylamide toxicity using a green paramecium. Toxicol In Vitro 19(1):99–105. doi:10.1016/j.tiv.2004.06.012

Tanaka H, Asano T, Schroeder ED, Tchobanoglous G (1998) Estimating the safety of wastewater reclamation and reuse using enteric virus monitoring data. Water Environ Res 70(1):39–51. doi:10.2175/106143098X126874

Tang JYM, Busetti F, Charrois JWA, Escher BI (2014) Which chemicals drive biological effects in wastewater and recycled water? Water Res 60:289–299. doi:10.1016/j.watres.2014.04.043

Tansel B (2008) New technologies for water and wastewater treatment: a survey of recent patents. Recent Pat Chem Eng 1:17–26. doi:10.2174/2211334710801010017

Tchobanoglous G, Burton FL, Stensel HD (2003) Wastewater engineering: treatment and ruse, 4th edn. McGraw-Hill, New York

Vargas VMF, Migliavacca SB, de Melo AC, Horn RC, Guidobono RR, de Sá Ferreira ICF, Pestana MHD (2001) Genotoxicity assessment in aquatic environments under the influence of heavy metals and organic contaminants. Mutat Res-Gen Tox 490(2):141–158. doi:10.1016/S1383-5718(00)00159-5

Verhaert V, Covaci A, Bouillon S, Abrantes K, Musibono D, Bervoets L, Verheyen E, Blust R (2013) Baseline levels and trophic transfer of persistent organic pollutants in sediments and biota from the Congo River Basin (DR Congo). Environ Int 59:290–302. doi:10.1016/j.envint.2013.05.015

Vinje J, Hamidjaja RA, Sobsey MD (2004) Development and application of a capsid VP1 (region D) based reverse transcription PCR assay for genotyping of genogroup I and II noroviruses. J Virol Methods 116:109–117. doi:10.1081/SAC-200055715

Vittozzi L, De Angelis G (1991) A critical review of comparative acute toxicity data on freshwater fish. Aquat Toxicol 19(3):167–204. doi:10.1016/0166-445X(91)90017-4

Voua Otomo P, Wepener V, Maboeta MS (2014) Single and mixture toxicity of gold nanoparticles and gold (III) to Enchytraeus buchholzi (Oligochaeta). Appl Soil Ecol 84:231–234. doi:10.1016/j.apsoil.2014.08.007

Wahsha M, Bini C, Argese E, Minello F, Fontana S, Wahsheh H (2012) Heavy metals accumulation in willows growing on Spolic Technosols from the abandoned Imperina Valley mine in Italy. J Geochem Explor 123(0):19–24. dio:10.1016/j.gexplo.2012.07.004

Walter MV, Vennes JW (1985) Occurrence of multiple-antibiotic-resistant enteric bacteria in domestic sewage and oxidation lagoons. Appl Environ Microbiol 50:930–933

Wang W (1987) Factors affecting metal toxicity to (and accumulation by) aquatic organisms—overview. Environ Int 13(6):437–457. doi:10.1016/0160-4120(87)90006-7

Wang XC, Qiu FG, Xue XP, Lu X (2005) Application of a virological tracer method for the assessment of pathogen removal by physicochemical treatment and chemical disinfection. Water Sci Technol 52(8):205–212

Wang LJ, Liu SS, Yuan J, Liu HL (2011) Remarkable hormesis induced by 1-ethyl-3-methyl imidazolium tetrafluoroborate on Vibrio qinghaiensis sp.-Q67. Chemosphere 84(10):1440–1445. doi:10.1016/j.chemosphere.2011.04.049

Ward RL, Bernstein DL, Young EC, Sherwood JR, Knowlton DR, Schiff GM (1986) Human rotavirus studies in volunteers: determination of infectious dose and serological responses to infection. J Infect Dis 154(5):871

Watson K, Shaw G, Leusch FDL, Knight NL (2012) Chlorine disinfection by-products in wastewater effluent: bioassay-based assessment of toxicological impact. Water Res 46(18):6069–6083. doi:10.1016/j.watres.2012.08.026

Wei D, Tan Z, Du Y (2011) A biological safety evaluation on reclaimed water reused as scenic water using a bioassay battery. J Environ Sci 23(10):1611–1618. doi:10.1016/S1001-0742(10)60631-6

Wheeler JR, Panter GH, Weltje L, Thorpe KL (2013) Test concentration setting for fish in vivo endocrine screening assays. Chemosphere 92(9):1067–1076. doi:10.1016/j.chemosphere.2013.01.102

WHO (2004) Guidelines for drinking-water quality, 3rd edn. World Health Organization, Geneva, Switzerland

Wibbertmann A, Mangelsdorf I, Gamon K, Sedlak R (2011) Toxicological properties and risk assessment of the anionic surfactants category: alkyl sulfates, primary alkane sulfonates, and α-olefin sulfonates. Ecotoxicol Environ Saf 74(5):1089–1106. doi:10.1016/j.ecoenv.2011.02.007

Wu MN, Wang XC, Ma XY (2013) Characteristics of THMFP increase in secondary effluent and its potential toxicity. J Hazard Mater 261:325–331. doi:10.1016/j.jhazmat.2013.07.022

Xu J, Wu L, Chang AC, Zhang Y (2010) Impact of long-term reclaimed wastewater irrigation on agricultural soils: a preliminary assessment. J Hazard Mater 183(1–3):780–786. doi:10.1016/j.jhazmat.2010.07.094

Ye Z, Zhao Q, Zhang M, Gao Y (2011) Acute toxicity evaluation of explosive wastewater by bacterial bioluminescence assays using a freshwater luminescent bacterium, Vibrio qinghaiensis sp. Nov. J Hazard Mater 186(2–3):1351–1354. doi:10.1016/j.jhazmat.2010.12.013

Yi H, Si L (2007) Vicia root-mirconucleus and sister chromatid exchange assays on the genotoxicity of selenium compounds. Mutat Res-Gen Tox 630(1–2):92–96. doi:10.1016/j.mrgentox.2007.03.003

Yi L, Jiao W, Chen X, Chen W (2011) An overview of reclaimed water reuse in China. J Environ Sci 23(10):1585–1593. doi:10.1016/S1001-0742(10)60627-4

Yoon Y, Ryu J, Oh J, Choi BG, Snyder SA (2010) Occurrence of endocrine disrupting compounds, pharmaceuticals, and personal care products in the Han River (Seoul, South Korea). Sci Total Environ 408(3):636–643. doi:10.1016/j.scitotenv.2009.10.049

Zenk MH (1996) Heavy metal detoxification in higher plants—a review. Gene 179(1):21–30. doi:10.1016/S0378-1119(96)00422-2

Zhang CM, Wang XC (2012) Health risk assessment of urban surface waters based on real-time PCR detection of typical pathogens. Hum Ecol Risk Assess 18(2):329–337. doi:10.1080/10807039.2012.650585

Zhang CM, Wang XC (2014) Distribution of enteric pathogens in wastewater secondary effluent and safety analysis for urban water reuse. Hum Ecol Risk Assess 20(3):797–806. doi:10.1080/10807039.2013.810957

Zhang XJ, Wang ZS, Gu XS (1991) Simple combination of biodegradation and carbon adsorption-the mechanism of the biological activated carbon process. Water Res 25(2):165–172. doi:10.1016/0043-1354(91)90025-L

Zhang CM, Liu YJ, Wang XC, Xue XP (2007) Study on membrane adsorption-elution method for concentration of enteroviruses from environmental waters. Environ Sci 28(7):1543–1547. doi:10.3321/j.issn:0250-3301.2007.07.024 (in Chinese with English abstract)

Zhang YH, Liu SS, Song XQ, Ge HL (2008) Prediction for the mixture toxicity of six organophosphorus pesticides to the luminescent bacterium Q67. Ecotoxicol Environ Saf 71(3):880–888. doi:10.1016/j.ecoenv.2008.01.014

Zhang J, Liu SS, Liu HL (2009a) Effect of ionic liquid on the toxicity of pesticide to Vibrio-qinghaiensis sp.-Q67. J Hazard Mater 170:920–927. doi:10.1016/j.jhazmat.2009.05.056

Zhang Y, Marrs CF, Simon C, Xi C (2009b) Wastewater treatment contributes to selective increase of antibiotic resistance among acinetobacter spp. Sci Total Environ 407(12):3702–3706. doi:10.1016/j.scitotenv.2009.02.013

Zhang CM, Wang XC, Liu YJ, Peng DC (2010) Simultaneous detection of enteroviruses from surface waters by real-time RT-PCR with universal primers. J Environ Sci 22(8):1261–1266. doi:10.1016/s1001-0742(09)60248-5

Zhang J, Liu SS, Yu ZY, Liu HL (2012a) Significant contributions of ionic liquids containing tetrafluoroborate and trifluoromethanesulfonate to antagonisms and synergisms in multi-component mixtures. J Hazard Mater 209–210:158–163. doi:10.1016/j.jhazmat.2012.01.003

Zhang CM, Wang XC, Zhou JH, Ji Z (2012b) Comparative study on enteric pathogens and their relations with fecal indicator bacteria in urban surface waters. Acta Sci Circum 32(11):2789–2794 (in Chinese with English abstract)

Zhang K, Wei YL, Zeng EY (2013a) A review of environmental and human exposure to persistent organic pollutants in the Pearl River Delta, South China. Sci Total Environ 463–464:1093–1110. doi:10.1016/j.scitotenv.2012.10.104

Zhang CM, Wang XC, Li X, Zhao LF, Zeng S (2013b) Quantification and distribution of Salmonella spp. and Salmonella typhi in wastewater treatment plants. J Pure Appl Microbiol 7(1):409–415

Zhang CM, Miao YH, Zhang QP, Xu H (2014) Effects of chlorination and ultraviolet disinfection on antibiotic resistance of fecal coliforms in secondary settling tank effluent of municipal wastewater treatment plant. Res Environ Sci 27(4):422–426. doi:10.13198/j.issn.1001-6929.2014.04.11 (in Chinese with English abstract)

Zhou QX, Kong FX, Zhu L (2004). Ecotoxicology. Science Press, Beijing (in Chinese)

Zhu XW, Liu SS, Ge HL, Liu Y (2009) Comparison between the short-term and the long-term toxicity of six triazine herbicides on photobacteria Q67. Water Res 43(6):1731–1739. doi:10.1016/j.watres.2009.01.004

Chapter 4
A Real Case of Water Reuse Through a Water Cycle

Abstract This chapter introduces a novel model case of water cycle management for a semi-decentralized water and wastewater system with reclaimed water use in a university campus in Xi'an, China where less than 3,000 m^3/d of groundwater is the only freshwater source for up to 35,000 population and 80 ha area with 60 % green coverage. Under the constraints of limited source water and large water demand for potable and non-potable purposes, the water system was designed by integrating water supply, sewerage, water reclamation and reuse, and water landscaping in one framework, utilizing the landscaping lake as a buffer zone in a mimicking natural water cycle with a high recycling ratio, and combining engineering and natural processes for water quality control. With the freshwater used only for potable purposes and the reclaimed water for all non-potable uses, the water application efficiency was doubled and safety water use was secured.

Keywords Water reuse · Water cycle · Landscape lake · Natural purification

4.1 Case Description

4.1.1 Background of the Project

The real case to be discussed is a project of water reuse in a university campus located in the eastern suburb of Xi'an city, Shaanxi Province in northwestern China. As shown in Fig. 4.1, annual precipitation is unevenly distributed in China. In the southern basins, annual precipitation can be as high as 2,000 mm or more, while in the northern basins, it is usually lower than 800 mm and even less than 100 mm to the northwestern side. As a result, the quantity ratio of water use to its renewable resource is 13.9 % in the southern basins, contrasting to 50.0 % in the northern basins, according to the latest data (Ministry of Water Resources 2012). Therefore, northern China has been suffering from serious water shortage for a long time,

X.C. Wang et al., *Water Cycle Management*,
SpringerBriefs in Water Science and Technology,
DOI 10.1007/978-3-662-45821-1_4

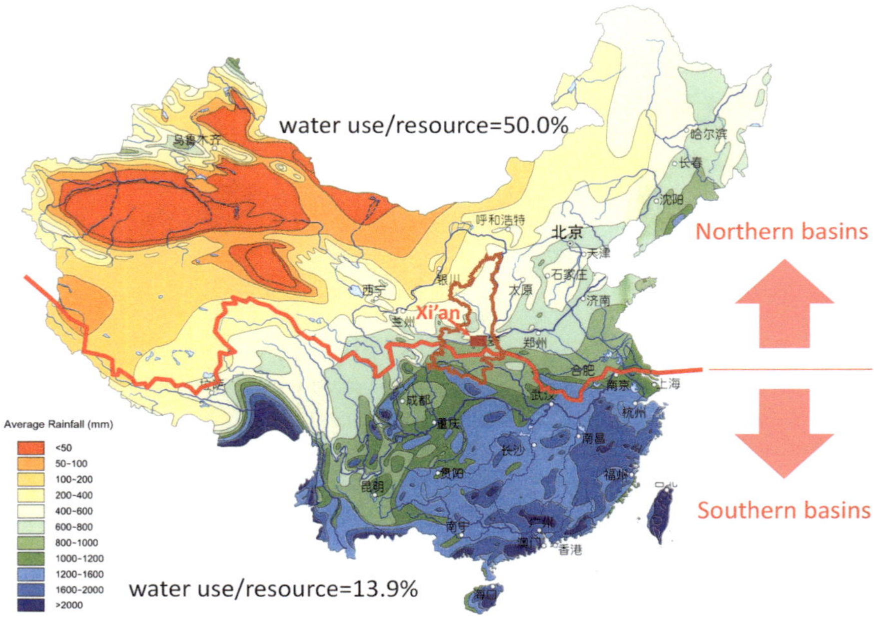

Fig. 4.1 Distribution of annual precipitation in China and water use to resource ratio in northern and southern basins. Precipitation map based on long-term average, provided by Department of Water Resources, China Institute of Water Resources and Hydropower Research; graph by Wang XC

especially since the 1980s when fast economic development and urbanization began in the whole country.

Xi'an, where the project is located, is the capital city of Shaanxi Province. This province consists of both areas in the southern and northern basins. As shown in more detail in Fig. 4.2, the northern part of the province belongs to the Yellow River Basin while the southern part belongs to the Yangtze River Basin. Although the total volume of the renewable resources accounted to 39.05 billion m^3 in 2012 and with an average per capita amount as 1041.9 m^3, which was about half of the national average (2186.1 m^3/person), as the northern part takes 68.5 % of the provincial territory and 77.3 % of the population but only 21.1 % of the renewable water resource, it results in a per capita water resource in the northern part of the province of only 284.4 m^3, which was only 13 % of the national average and much lower than the level of the worldwide recognized absolute water scarcity (Wang and Jin 2006). Therefore, the northern part of Shaanxi Province, including Xi'an City is among the most water deficient region in China.

Under such a condition, the Chinese government put forward a policy that treated wastewater has to be reused up to 22.8 % of the total quantity of the treated domestic wastewater by 2015 (The State Council 2012). For Shaanxi Province, the quantity ratio of water reuse to wastewater treatment was planned to be as high as about 49 % to the same target year to mitigate the water shortage problem. To reach

Fig. 4.2 Map of Shaanxi Province and its river basins. Base map provided by Shaanxi Administration of Surveying, Mapping and Geoinformation: map no. S(2012)008; graph by Wang XC

such a goal, the installation of both centralized and decentralized wastewater treatment and reuse systems is required. A centralized wastewater treatment and reuse system is often an extension of the conventional urban wastewater system characterized by large-scale collection and treatment, plus a large-scale network to transfer and distribute the reclaimed water to various users. In contrast to this, a decentralized system is characterized by onsite treatment and reuse, so that the construction of long pipelines for wastewater collection and reclaimed water distribution may not be required (Chen and Wang 2009; Mankad and Tapsuwan 2011; Marlow et al. 2013). In addition to this, a decentralized system can be designed and constructed in more flexible ways, according to the practical demands for reclaimed water supply and the actual situation in the service area, so that water cycle management can be well realized for high efficient water reclamation and reuse.

4.1.2 Project Outline

The project to be introduced is a wastewater treatment/reclamation and reuse system designed and installed in Xi'an Siyuan University. As shown in Fig. 4.3, this university is located in the southeast suburban area of Xi'an city on a hill elevated about 200 m higher than the surroundings. At the time, when the university was established in 2001, the urban area of Xi'an city was not as large as it is at present and neither urban water supply nor sewerage systems covered this remote area. Therefore, the university was permitted to dig five deep wells for potable and non-potable water supply through an independent system. Regarding the used water, an activated sludge treatment unit was constructed for meeting the regulation of discharge down to the hill. Part of the treated wastewater was occasionally used for irrigating trees and grassland in the vicinity. The university was among the first group of private universities established in the boom of higher education development at the turn of the century. The teaching staff mostly works concurrently, so that the population for water supply is mainly students living in the dormitories provided by the university.

The maximum capacity of water supply from the five groundwater wells is up to 3,000 m^3/d, which could meet the demand for potable and non-potable water uses in the earlier days when the student number was not so large. However, along with the fast development of higher education in China in the following years, the scale of the university was expanded rapidly and began to face a serious water shortage problem. At present, the number of students living in the campus amounts to about 25,000, and the campus covers an area about 80 ha of which about 50 ha (60 % of the area) are green belts. It is also anticipated that the student number will increase to about 35,000 in 5 years. By a simple calculation, the current available water source is only 120 L/d per capita and will be decreased to 85 L/d per capita after 5 years, which can marginally meet the lowest requirement for potable consumption but definitely insufficient for other non-potable consumptions, because only to

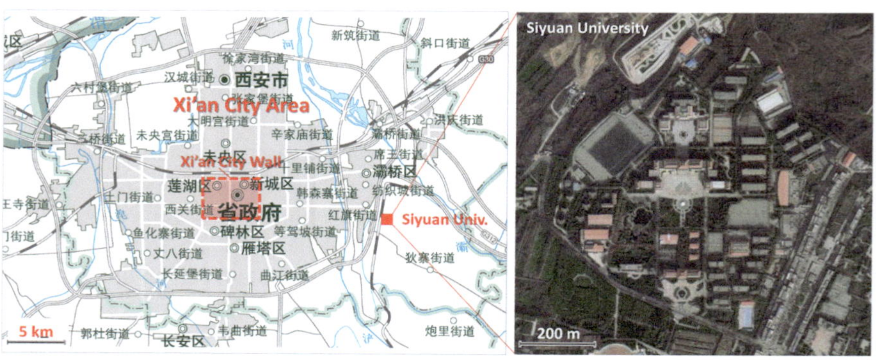

Fig. 4.3 Project location. Base map provided by Shaanxi Administration of Surveying, Mapping and Geoinformation: map no. S(2012)008; graph by Wang XC

irrigate the large green belt will require a large amount of water. Facing the envisaged problem for water supply, the university has to seek countermeasures either by applying to the municipal government for access to the urban water supply network or to practice on campus wastewater treatment and non-potable water reuse. As water reclamation and reuse is encouraged and supported by local authorities, the university has chosen the option of water reclamation and reuse to solve the present and future problems of water shortage.

4.2 System Design and Implementation

4.2.1 *Water Demand and Availability*

Water demand and availability were evaluated by water budget analysis, basically following the quantitative analysis model described in Sect. 2.3.2. The relationships between water demands, available water sources, and categories of water supply in the university campus can be depicted in Fig. 4.4, in which D_1, D_2, D_3, D_4, D_5 are water demands for direct drinking, bathing/washing, toilet flushing, gardening, and waterfront landscaping, respectively, and F_0 and R_0 are sources from groundwater and collectable used-water, respectively, while F_1 and R_1 are quantities of fresh water supply and reclaimed water supply, respectively, all as m^3/d. Under a consideration of the limited amount of the groundwater for potable water supply, it should be preferably used for direct drinking (including cooking) and bathing/washing. Therefore, the water budget relationship for groundwater supply is:

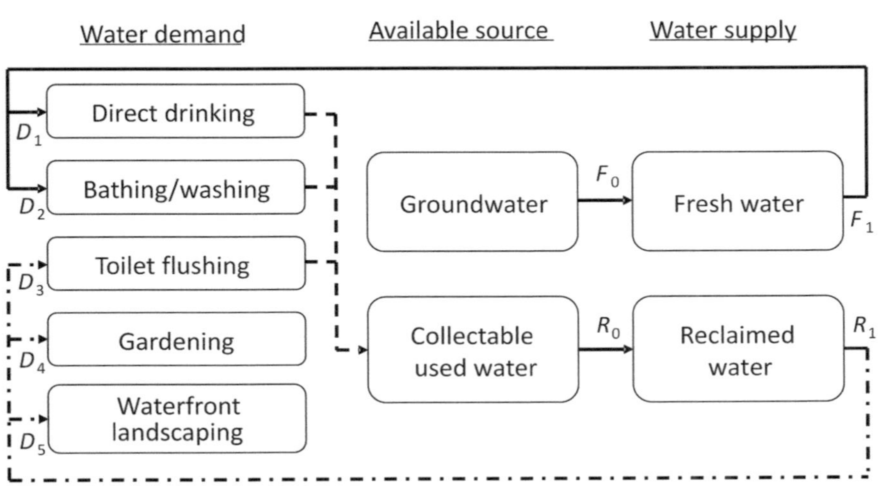

Fig. 4.4 Relationship between water demands, available sources, and categories of water supply. Graph by Wang XC and Luo L

$$F_0 \geq F_1 = D_1 + D_2 \tag{4.1}$$

The reclaimed water should be used for all non-potable purposes including toilet flushing, gardening and waterfront landscaping. Therefore, the water budget relationship for reclaimed water supply is:

$$R_0 \geq R_1 = D_3 + D_4 + D_5 \tag{4.2}$$

On the other hand, the source water for water reclamation should be the collectable used-water including those after drinking, bathing/washing, and toilet flushing. Therefore the water budget relationship for used-water collection is:

$$R_0 = D_1 + D_2 + D_3 - (\mathit{water\,loss\,during\,water\,use}) \tag{4.3}$$

Following the above water budget relationships, the first question is whether or not the groundwater supply can meet the needs for potable water use. By a survey on the service level and water consumptions for potable uses at various locations such as washing at the students dormitories, bathing or showering at the bath center (very common in China for collective life on campus), students canteens, and educational facilities that needed fresh water supply), it was estimated that the per capita water consumption only for potable purposes was between 70 and 80 L/d. By taking 80 L/d as the demand for per capita potable water use, the current total water demand for potable water supply would be 2,000 m^3/d (student number as 25,000) while that for the future would be 2,800 m^3/d (student number increases to 35,000). The conclusion is that the maximum capacity of groundwater withdrawal of 3,000 m^3/d can meet the current and future needs for potable water supply while all non-potable uses are definitely impossible to be covered by groundwater supply.

The second question is whether or not water reclamation can meet the demands for all non-potable water uses. To answer this question, a projection can be conducted as shown in Table 4.1 for various purposes of non-potable water supply. The projection is based on the assumption that in the near future there will be an increase of student number from 25,000 to 35,000 while the campus area will remain as it is currently. Considering the categories of non-potable water use for toilet flushing, gardening, lake replenishment, and water loss from the lake surface, the change of water demand from now to the future will only be that for toilet flushing, which is calculated on per capita basis. As a result (Table 4.1), it is projected that the total non-potable water demand at current time is 3,450 m^3/d, while that for the future will be 3,750 m^3/d. Either of these two figures is apparently much higher than the original capacity of groundwater supply (3,000 m^3/d). This implies that only when the reclaimed water is to be used more than once can the various demands for non-potable water use be met.

Therefore, there comes the third question on how to realize water reclamation and reuse in a more efficient way. In Table 4.1, the requirement for 20 % daily replacement of water in the landscaping lake is for controlling the hydraulic retention time (HRT), so that water stagnation or eutrophication can be easily prevented. The

Table 4.1 Projection of non-potable water demands

	Water use	Specification	Quantity	Demand	Remarks
1	Toilet flushing	30 L/d/person[a]	25,000–35,000 persons	750–1,050 m^3/d	
2	Gardening	3 L/m^2/d[b]	500,000 m^2	1,500 m^3/d	For 50 ha green belt
3	Lake replenishment[c]	20 % daily replacement	50,000 m^3	1,000 m^3	For lakes of 5,000 m^2 and 1 m depth
4	Lake water loss[d]	20 % of lake replenishment	1,000 m^3	200 m^3	Evaporation and other loss
5	Total			3,450–3,750 m^3/d	Including item 3
				2,450–2,750 m^3/d	Excluding item 3

[a] According to Chinese specification of 20–40 L/d/person for toilet flushing (Li 2002)
[b] According to Chinese specification of 2.5–3.5 L/m^3 /d for gardening (Li 2002)
[c] According to Chinese specification of water replacement per time in 4–5 days interval (Li 2002)
[d] According to local experience

water demand of 1,000 m^3/d for this purpose may not physically consume water at all. Following the concept of water cycle management described in Chap. 2, it is possible to take the landscaping lake as an element or subsystem in the water cycle but not merely a reclaimed water user. In addition to its function for waterfront landscaping, if the storage capacity of the lake is utilized in the water cycle, then lake replenishment can be fulfilled by just letting the reclaimed water flow in, be stored in the lake with a given HRT, and finally be pumped out for other non-potable uses such as toilet flushing and gardening, which demand much larger quantities of water than that for lake replenishment (Table 4.1). When the demand for lake replenishment is no longer taken as water consumption, the total non-potable water demand will be greatly decreased as shown in the last row of Table 4.1.

4.2.2 System Design

Following the water budget relationships discussed in Sect. 4.2.1, the water system was configured in a form as shown in Fig. 4.5 by taking into account the future condition of groundwater withdrawal of 2,800 m^3/d for potable water supply to 3,5000 persons with a per capita potable water consumption as 80 L/d. After potable use, the collectable used-water was supposed to be 80 % of the supply, namely 2,240 m^3/d according to local experience. In order to enlarge the collectable amount of the used-water as the source for water reclamation, all the reclaimed water after being used for toilet flushing were also collected and added to the inflow of the treatment/reclamation plant. The added inflow was estimated as 90 % of the water for toilet flushing at 945 m^3/d, so that the total inflow would become 3,185 m^3/d. If 90 % of the inflow could be reclaimed for non-potable water supply, the potential of reclaimed water production would become 2,867 m^3/d, which is larger than the total

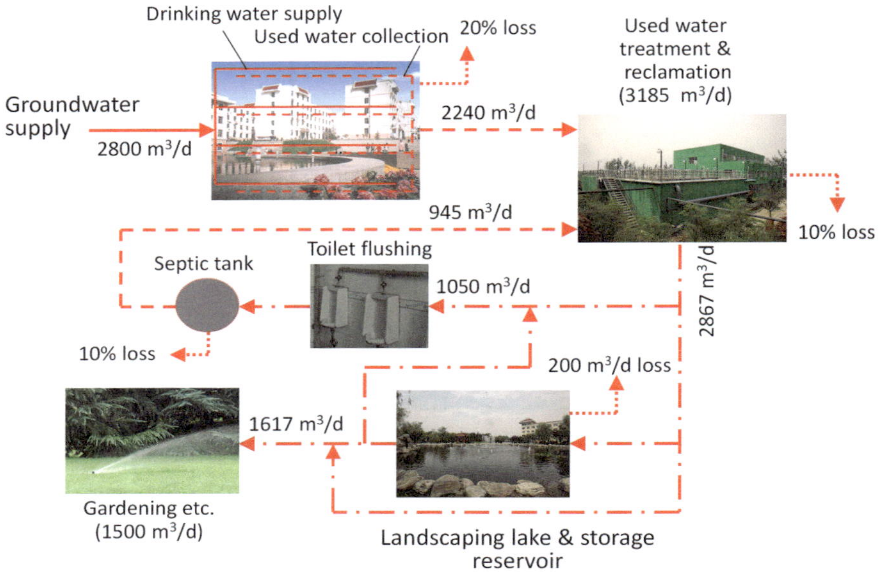

Fig. 4.5 Configuration of the on campus water system. Photo images and graph by Wang XC

non-potable water demand of 2,750 m^3/d shown in Table 4.1, indicating the feasibility of source enlargement through an inner cycle within the system. According to the distribution of points for reclaimed water use such as toilets, green belts for gardening/irrigation and so on, about a half quantity of the reclaimed water would be sent directly to the points of use while another half quantity was led to a landscaping lake located in the central courtyard of the campus. With a storage capacity of about 5,000 m^3, the landscaping lake is also performing the function of a storage or regulation reservoir, which receives reclaimed water from the treatment/reclamation plant at its inlet while providing stored water for various uses by pumping from its outlet. Lake water replacement or replenishment can thus, be dynamically conducted without physical consumption other than evaporation or seepage, which have already been taken into account in Table 4.1.

From a quantitative viewpoint, through the water cycle shown in Fig. 4.5, water source enlargement can be realized. The total available water quantity includes the 2,800 m^3/d of groundwater for potable supply, and the same or a little bit larger amount of reclaimed water for non-potable supply, so that the efficiency of water application can be higher than 200 %.

4.2.3 Process Selection for Water Treatment/Reclamation

Selection of treatment processes for water reclamation should principally meet the requirement for pollutant removal from the collected used-water so that the reclaimed water can meet a quality suitable for reuse. As discussed in Sect. 2.3.3, the required removal of the mass of pollutant i can be determined by materials balance analysis. Referring to the water system shown in Fig. 4.5, a materials balance relationship can be drawn regarding pollutant i as shown in Fig. 4.6. It is noticeable that as the water after toilet flushing returns to the treatment/reclamation plant, there is an inner loop in the system with part of the pollutants circulating. The water uncollectable after use includes those for gardening and lake water loss.

By materials balance calculation, it can be taken that $\Delta M_{i,2}$, the mass of pollutant i to be removed in the used-water treatment/reclamation processes should be:

$$\Delta M_{i,2} = F_1 C_{i,1} + \Delta M_{i,1} + \Delta M_{i,3} - \Delta M_{i,4} - (R_1 - D_3) C_{i,2} \tag{4.4}$$

where all parameters are as indicated in Fig. 4.6. Because $C_{i,2}$ is the concentration of substance i in the reclaimed water, which should meet the requirement for the reuse purposes, Chinese standards on water reuse (Standardization Administration of China 2002a, b) are referred for selecting appropriate $C_{i,2}$ regarding suspended matter (SS or turbidity), organic substance (BOD), salinity (TDS), nutrients (TP, TN, and NH_3-N), and microorganism (fecal coliform).

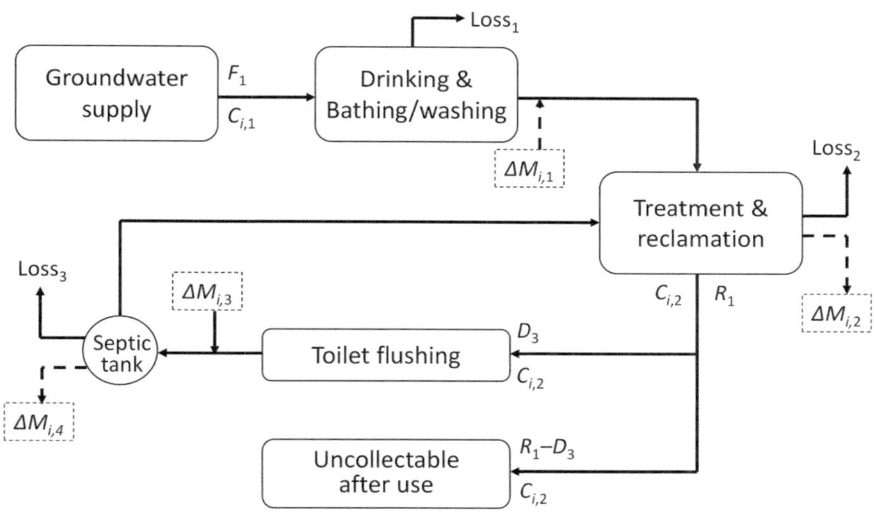

Fig. 4.6 Materials balance relationship regarding pollutant i in the campus water system. F_1, R_1 and D_3: quantities of groundwater supply, total reclaimed water supply and reclaimed water for toilet flushing, respectively, all as m^3/d; $C_{i,1}$ and $C_{i,2}$: concentrations of pollutant i in groundwater and reclaimed water, respectively, all as mg/L or g/m^3; $Loss_1$, $Loss_2$ and $Loss_3$: quantities of water loss during potable use, treatment/reclamation and septic tank desludge, respectively, all as m^3/d; $\Delta M_{i,1}$, $\Delta M_{i,2}$, $\Delta M_{i,3}$, and $\Delta M_{i,4}$: masses of pollutant i entered the water after potable use, removed by treatment, entered the water after toilet flushing, and removed from septic tank, respectively, all as g/d. Graph by Wang XC and Luo L

As shown in Table 4.2, considering the main purposes of water reuse, in this case for toilet flushing, replenishment of landscaping lake, and gardening, higher water quality requirements are specified for the landscaping lake regarding BOD, TP, TN, and NH_3-N, while requirements are higher for toilet flushing regarding turbidity and fecal coliform. For gardening, the requirement is higher for TDS. In order to produce the reclaimed water with a quality to meet all these uses, the target values, namely $C_{i,2}$, for these parameters are set as the italic figures in Table 4.2.

On the other hand, the total mass of pollutant i entering the water system as a result of various water uses can be calculated as:

$$TM_{i,in} = F_1C_{i,1} + \Delta M_{i,1} + \Delta M_{i,3} + D_3C_{i,2} \tag{4.5}$$

and the mass of pollutant i residual in the reclaimed water can be evaluated as:

$$TM_{i,out} = R_1C_{i,2} \tag{4.6}$$

Therefore, the required removal of pollutant i is:

$$TR_i\,(\%) = [(TM_{i,in} - TM_{i,out})/TM_{i,in}] \times 100\,\% \tag{4.7}$$

where TR_i is total removal by the whole process of treatment including septic tanks for the pretreatment of feces (Fig. 4.6).

The calculation of $M_{i,out}$ using Eq. (4.6) can be easily done by taking the total reclaimed water production as R_1 and the target water quality $C_{i,2}$ as those indicated in Table 4.2, but the calculation of $M_{i,in}$ using Eq. (4.5) needs an evaluation of the pollutant loading for obtaining $\Delta M_{i,1}$, $\Delta M_{i,3}$, and $\Delta M_{i,4}$. Table 4.3 shows the estimated daily loadings for organics (COD and BOD) and nutrients (TN and TP) based on their unit loadings according to experiences in China (Wang et al. 2008; Feng et al. 2009).

Using the loadings for BOD, TN, and TP listed in Table 4.3 to calculate the corresponding removals required following Eqs. (4.5), (4.6) and (4.7), the results are obtained as shown in Table 4.4. The groundwater supplied for potable use is almost free from BOD and any type of phosphorus, based on long-term water

Table 4.2 Chinese water quality criteria for different reuse purposes

Parameter	Unit	Reuse purpose		
		Toilet flushing	Landscaping lake	Gardening
BOD	mg/L	10	*6*	
SS (Turbidity[a])	mg/L (NTU)	*5*[a]	10	10*
TDS	mg/L	1,500	–	*1,000*
TP	mg/L	–	*0.5*	–
TN	mg/L	–	*15*	–
NH_3-N	mg/L	10	*5*	20
Fecal coliform	1/L	*3*	2,000	3

[a] Turbidity values

Table 4.3 Estimation of organic and nutrients loading

Parameter	Unit loading[a] (g/d/person)	Population (persons)	Calculated loading (kg/d)[b]		
			Total	Fecal	Miscellaneous
COD	42		1470.0	558.6	911.4
BOD[c]	25.2		882.0	335.2	546.8
TN	14	35,000	490.0	401.8	88.2
TP	0.8		28.00	19.04	8.96

[a] Unit loadings according to experiences in China regarding COD, TN, and TP (Wang et al. 2008; Feng et al. 2009)
[b] Percent of each of the fecal and miscellaneous loadings estimated as 38 and 62 % for COD, 82 and 18 % for TN, 68 and 32 % for TP (Feng et al. 2009)
[c] BOD/COD ratio as 0.6 in wastewater from domestic source according to local experience

Table 4.4 Calculation of the required BOD, TN and TP removals

Parameter	BOD	TN	TP	Remarks
F_1 (m^3/d)		2,800		
R_1 (m^3/d)		2,867		
D_3 (m^3/d)		1,050		
$C_{i,1}$ (mg/L)	0	7.8	0	TN as NO_3^{-1} in groundwater
$C_{i,2}$ (mg/L)	6	15	0.5	
$\Delta M_{i,1}$ (kg/d)	546.8	88.2	8.96	
$\Delta M_{i,3}$ (kg/d)	335.2	401.8	19.04	
$TM_{i,in}$ (kg/d)	888.3	527.6	28.53	Equation (4.5)
$TM_{i,out}$ (kg/d)	17.2	43.0	1.43	Equation (4.6)
TR_i (%)	*98.1*	*91.8*	*95.0*	Equation (4.7)

quality monitoring data. However, it contains some nitrate (NO_3^{-1}) to an average concentration of 7.8 mg/L. This has been taken into account in the calculation of TN removal as indicated in Table 4.3. It can be seen that for controlling the reclaimed water quality to meet the requirement for landscaping water use (Table 4.2), the treatment process should be designed to achieve very high removals of BOD, TN and TP. Therefore, a sophisticated process as shown in Fig. 4.7 has been selected as a combination of biological unit in anaerobic-anoxic-oxic (A^2O) array for effective nitrogen and phosphorus removal followed by a membrane bioreactor (MBR) for enhancing both biological treatment and solid/liquid separation. In order to achieve higher phosphorus removal, chemical precipitation (using polyaluminium chloride, PAC as chemical) has also been employed before the MBR unit. Although septic tanks can assist pollutants removal in the system (Fig. 4.6), in order to preserve the carbon source for effective denitrification in the A^2O unit, their hydraulic retention time is controlled as short as possible. The full capacity of the treatment facilities has been designed as 4,000 m^3/d to meet the future needs of system expansion.

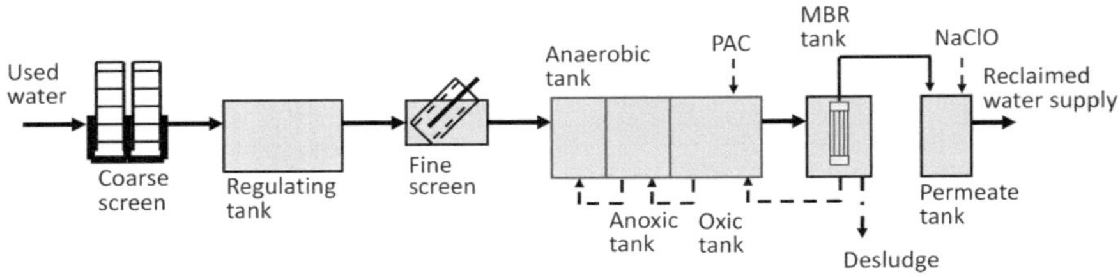

Fig. 4.7 Schematic diagram of the used-water treatment and reclamation process. Full capacity of treatment as 4,000 m^3/d, graph by Wang XC

In Table 4.2, SS or turbidity is also a water quality parameter to be controlled. However, with the MBR equipped with MF membranes, all suspended or turbid substances can be effectively removed. In the permeate tank, with NaClO injection for final disinfection the reclaimed water is free from fecal coliform and other pathogens. TDS is a parameter that cannot be controlled by such a treatment process. There may exist a risk of salt accumulation because of the recollection of the toilet flushing water for source enlargement. The condition of salt concentration of the reclaimed water will be discussed later in Sect. 4.3.

4.2.4 Landscaping Lake in the Water Cycle

As shown in Fig. 4.5, water reuse in the campus is through a water cycle, which starts from groundwater withdrawal from the wells for potable supply, and then enters a partially closed loop of used water collection, treatment and reclamation, reclaimed water distribution and reuse, and with a return of the collectable liquid to the treatment plant. A noticeable feature of the water cycle is the introduction of a lake, which performs both the functions of landscaping and water storage (Fig. 4.8). With its total storage volume up to 5,000 m^3, the lake provides sufficient capacity to absorb the quantitative fluctuation between reclaimed water production and its consumption for various uses. On the other hand, stepwise or escalated water use is realized in a manner shown in Fig. 4.8. Firstly, the reclaimed water with undisturbed quality is supplied directly to a series of water landscapes in the campus, which is the first-step water use, because the lake is downstream of their outlets to receive the outflows. Secondly, the lake itself is the major user of the reclaimed water for landscaping. This can be viewed as the second-step water use. Thirdly, downstream of the lake a pump station is provided for pumping the water stored in the lake to various locations for outdoor gardening and indoor toilet flushing, which is the third-step water use. It should be pointed out that in the first- and second-step water uses, water is not physically consumed except for a limited quantity of water loss due to evaporation. This provides a good condition for efficient use of the reclaimed water. In the third-step water use, water is virtually consumed by

Fig. 4.8 A landscaping lake in the water cycle. Photo images and graph by Wang XC

gardening while the water used for toilet flushing will be collected again to supplement the source for water reclamation.

The hydraulic retention time (HRT) of water in the lake is dynamically controlled through receiving the reclaimed water at its inlets and pumping water for various uses from its outlet. As a result, the HRT can be controlled within 5 days, which provides a good condition for preventing deterioration of the water quality due to water being stagnant, though the lake receives reclaimed water only. As the lake is for landscaping purposes in the central yard of the campus, various waterscape facilities, such as fountains, waterfalls, and water plants can also be utilized to assist aeration and enhance natural purification (Fig. 4.9).

4.3 Effects of Water Reuse Through a Water Cycle

4.3.1 Water Source Augmentation

Since the system was put into full operation in 2011, the problem of water shortage has been well solved. With most of the non-potable water uses shifted from groundwater supply to reclaimed water supply, as expected, the groundwater withdrawal has been controlled at about 2,000 m^3/d on average. Even in the summer season, groundwater supply does not exceed 2,500 m^3/d and over-pumping of groundwater for meeting the basic demands of various water uses no longer

Fig. 4.9 Enhancement of natural purification in the landscaping lake utilizing **a** rockwork with falling water, **b** fountains, **c** waterfalls, and **d** water plants. Photos by Wang XC

happens. At the current population level (about 25,000 students), the daily production of reclaimed water is about 2,500 m^3/d, which can meet the needs for toilet flushing and gardening in the whole campus area. As the lake can store a water volume for 2 days' non-potable water consumption, the contradiction between the constant reclaimed water production rate and timely fluctuation of water use has also been well solved.

The A^2O-MBR treatment system has been partitioned into several parallel trains to cope with the seasonal fluctuation of collectable used water as inflow to the treatment plant, especially the sudden decrease of inflow during the weekends and holidays. The lake also provides a buffer zone in the system to absorb such variation.

On the other hand, if Eq. (2.12) is used to evaluate the recycling ratio of water in the water cycle, referring to Fig. 4.4, it can be estimated that the fraction of water collected after potable water use is $\alpha = 0.8$, the fraction of water collected after reclaimed water use is $\beta = 0.33$, and the quantity ratio of reclaimed water to collected wastewater is $\gamma = 0.9$. It can thus, be calculated that under the current condition of water uses, the recycling ratio is:

$$\frac{R_1}{F_1} = \frac{\alpha\gamma}{1 - \beta\gamma} = 1.024 \tag{4.8}$$

indicating that, in the system, the quantity of total reclaimed water supply (R_1) is over the quantity of total potable water supply (F_1), and maximized water reclamation and reuse is realized.

4.3.2 Water Quality Aspects

The reclaimed water produced by the treatment process depicted in Fig. 4.7 has provided high quality water to meet the requirements for the reuse purposes shown in Table 4.2. Although in the process design, attention was mainly paid to BOD, TN, and TP control, with the high performance of the membrane for effective separation and the provision of final chlorination, the requirements for turbidity and fecal coliform have also been well met. In this section, discussions will not be on the treatment process itself, but the variation of water quality in the water cycle shown in Fig. 4.5, especially the landscaping lake (Fig. 4.8) because it performs the function of water storage and provides water for further reuse.

Figure 4.10 compares several conventional water quality parameters of the reclaimed water and the water stored in the landscaping lake based on long-term monitoring data. It can be seen that organic substances (COD) and ammonia-nitrogen (NH_3-N) concentrations in the lake water are slightly higher than those of the reclaimed water. This is believed to be due to non-point source contamination during water storage in the open lake because COD and NH_3-N can be from various sources relating to human activities. However, regarding total concentrations of nutrient salts, in terms of TN and TP, they tend to be lower in the lake than the reclaimed water, indicating that the lake can assimilate nutrient salts through natural processes, especially water plants.

Although total dissolved solids (TDS) is an important parameter relating to the suitability of the reclaimed water for various reuse purposes as shown in Table 4.2, we did not discuss it in Sect. 4.2.3 for the selection of treatment process because TDS cannot be removed by biological and/or physicochemical treatment, unless specific technologies such as electrodialysis or reverse osmosis (refer to Table 3.1) are applied. With domestic wastewater as the source for water reclamation, TDS is usually not a problem if the reclaimed water is used in an end-of-the-pipe manner, namely straightforward use without recollection. However, in the system shown in Fig. 4.5, the reclaimed water, after being used for toilet flushing is recollected and returned to the treatment plant for source enlargement. As one person may consume 8–10 g of salts daily, and the human body only absorbs a small portion of this, a great amount of salts are discharged with human wastes, especially in urine. Therefore, by a material balance calculation, similar to that described in Sect. 4.2.3, the accumulation of salts would inevitably occur, resulting in the elevation of TDS in the system

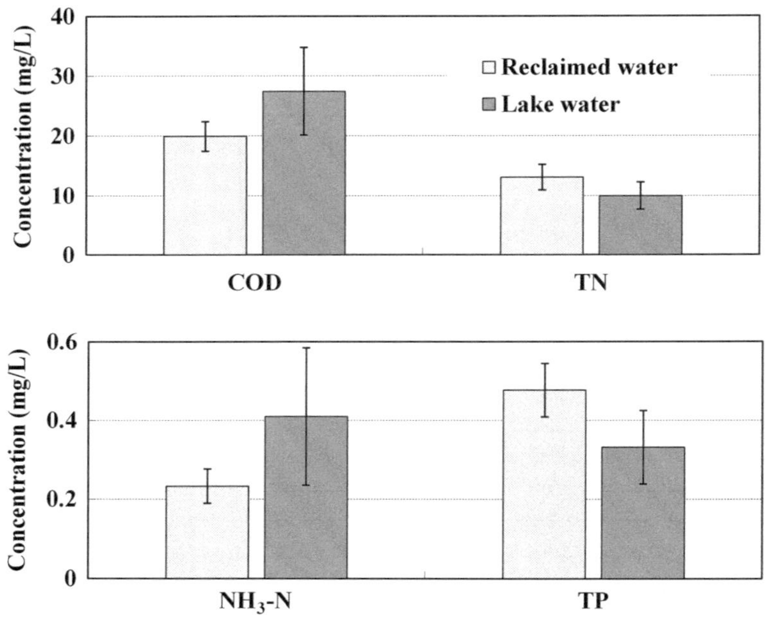

Fig. 4.10 Comparison of the reclaimed water and lake water for conventional water quality parameters. *Source* Unpublished data of the authors, graph by Wang XC

with long-term operation. This is, in fact, one issue of concern in designing such a water cycle. Nevertheless, as shown in Fig. 4.11, through a long-term monitoring of the reclaimed water and lake water regarding TDS, as well as electrical conductivity (EC), it has been found that comparing with the water from groundwater wells, which is the original source for campus water supply, TDS and EC of the reclaimed water are about 30 % elevated (in term of TDS from an average value of 320 mg/L in the groundwater to an average value of 430 mg/L in the lake water). However, the average TDS in the lake water is significantly reduced and returns to a level about 340 mg/L, which is just slightly higher than that of the groundwater. The reduction of salt content in the lake may possibly be due to several natural processes, such as uptake by aquatic vegetation (water plants and algae), adsorption by soil particles, chemical reactions, and occasional dilution by rainwater. The real mechanisms for these actions need detailed investigation in further studies.

From the viewpoint of ecological safety control, the lake is also performing important roles as shown in Fig. 4.12, regarding ecotoxicities of the reclaimed water and lake water measured by bioassays using Q67 luminescent bacteria and Chlorella, typical alga species. It can be seen that the reclaimed water, though treated by MBR process and with high quality to meet the requirement for reuse purposes, still shows certain toxicity on bacteria and detectable toxicity on alga. However, after storage in the lake, the inhibition values are effectively decreased, indicating a favorable ecological condition in the lake for ecotoxicity control.

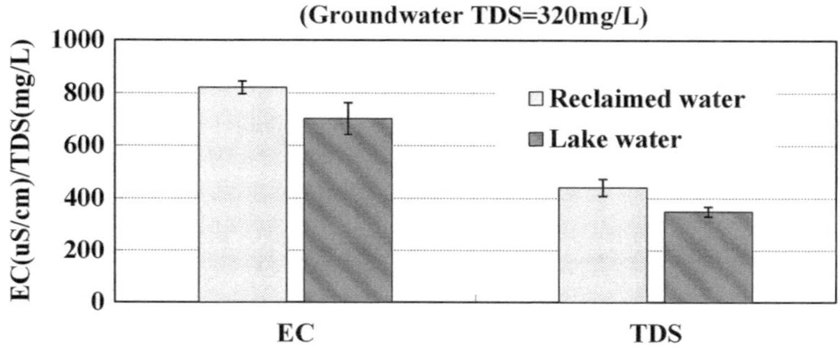

Fig. 4.11 Comparison of the reclaimed water and lake water for salt contents in terms of TDS and EC. *Source* Unpublished data of the authors, graph by Wang XC

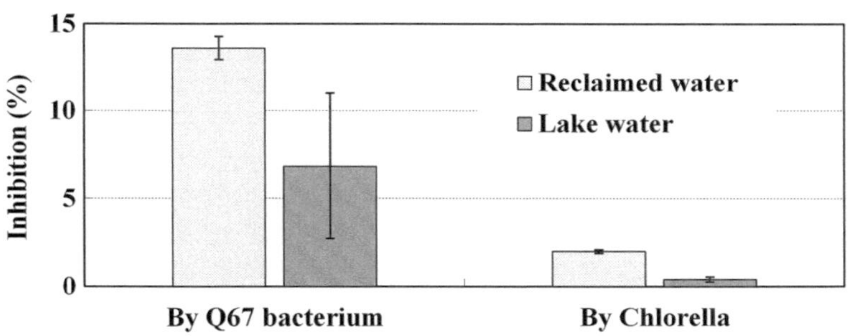

Fig. 4.12 Comparison of the reclaimed water and lake water for biotoxicities. *Source* Unpublished data of the authors, graph by Wang XC

The reclaimed water is free from bacterial and viral pathogens after membrane cutoff and final chlorination. Nevertheless, as the lake is an open water body located in the central yard of the campus and with the major waterfront landscape attracting visitors, the risk of secondary contamination with pathogens from nonpoint sources has to be evaluated. For this purpose, long-term monitoring of selected pathogens was conducted for typical bacteria such as *E. coli*, *Salmonella*, and *Sigella*, and viruses such as enteroviruses (EVs), rotaviruses, and noroviruses using real-time PCR techniques. Considering the influence of weather on the intrusion, growth and decay of pathogens in the lake, data were analyzed in three categories as sunny days, rainy days and storm periods. As shown in Fig. 4.13, almost all pathogens were detected from the lake with different frequencies and concentrations. In generally, bacteria were detected more frequently and with higher concentrations, especially for *E. coli*, which showed a concentration about 1×10^3 CFU/100 mL on average and as high as 1×10^4 CFU/100 mL in storm periods. Of the three kinds of viruses, EVs were detected more frequently and with higher concentrations (up to 1×10^2 CFU/100 mL) than rotaviruses and noroviruses. It is thus, identified that for the storage of the

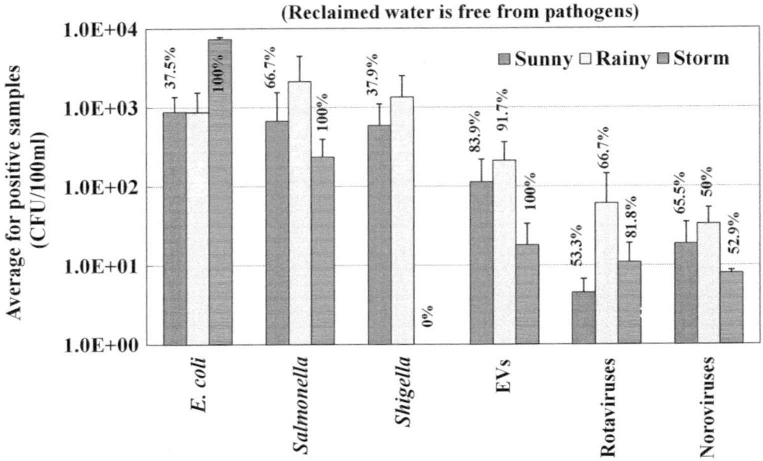

Fig. 4.13 Typical pathogenic bacteria and viruses detected from the lake water under different weather conditions. *Source* Unpublished data of the authors, graph by Wang XC

reclaimed water in open waters such as the landscaping lake, in this case, the pathogenic risk should be under concern. Onsite disinfection would be recommendable before the stored water is sent to various locations for toilet flushing and gardening.

4.3.3 Social and Environmental Benefits

The social and environmental benefits of water reclamation and reuse through a water cycle are obvious because the fresh water resource, namely the groundwater, is only used for potable purposes and the reclaimed water covers all non-potable water uses. As a result, the efficiency of water utilization is doubled, and the precious groundwater resource is saved. By the collection and reclamation of all collectable used waters, including the water after toilet flushing using the reclaimed water, wastewater discharge is almost completely eliminated, except for the overflow of excess water from the landscaping lake in rainy days. For a university with a large campus area and high green coverage with water landscapes, its maintenance needs plenty of water, but can be sustained by using the reclaimed water. The lake and the surrounding green belts have become the central water landscapes in the campus (Fig. 4.14). The mimicking of natural beauty has completely changed human perception on wastewater reuse.

Fig. 4.14 Scenery area in the central yard of the university with the landscaping lake and green belts nourished by reclaimed water. Photo provided by Xi'an Siyuan University

References

Chen R, Wang XC (2009) Cost–benefit evaluation of a decentralized water system for wastewater reuse and environmental protection. Water Sci Technol 59(8):1515–1522. doi:10.2166/wst.2009.156

Feng LP, Guo ZY, Xu ZC, Sun JW, Luo M, Tan ZY (2009) Preliminary study of TN and TP from human discharge (in Chinese). China Sci Technol Inform 1(1):16–17

Li GB (2002) Introduction to urban water engineering (in Chinese). China Architecture and Building Press, Beijing

Mankad A, Tapsuwan S (2011) Review of socio-economic drivers of community acceptance and adoption of decentralised water systems. J Environ Manage 92(3):380–391. doi:10.1016/j.jenvman.2010.10.037

Marlow DR, Moglia M, Cook S, Beale DJ (2013) Towards sustainable urban water management: a critical reassessment. Water Res 47(20):7150–7161. doi:10.1016/j.watres.2013.07.046

Ministry of Water Resources (2012) Report of water resources in 2012 (in Chinese). Ministry of Water Resources, People's Republic of China. http://www.mwr.gov.cn/zwzc/hygb/szygb/qgszygb/201405/t20140513_560838.html

Standardization Administration of China (2002a) Reuse of recycling water for urban: water quality standard for urban miscellaneous water consumption (GB/T 18920-2002, in Chinese). Standardization Administration, People's Republic of China

Standardization Administration of China (2002b) Reuse of recycling water for urban: water quality standard for scenic environment use (GB/T 18921-2002, in Chinese). Standardization Administration, People's Republic of China

The State Council (2012) Notice on the issuance of the 12th-five-year national plan for urban sewage treatment and reuse facilities construction (in Chinese). The State Council, People's Republic of China. http://www.gov.cn/zwgk/2012-05/04/content_2129670.htm

Wang XC, Jin PK (2006) Water shortage and needs for wastewater re-use in the north China. Water Sci Technol 53(9):35–44. doi:10.2166/wst.2006.267

Wang XC, Chen R, Zhang QH, Li K (2008) Optimized plan of centralized and decentralized wastewater reuse systems for housing development in the urban area of Xi'an, China. Water Sci Technol 58(5):969–975. doi:10.2166/wst.2008.456

Chapter 5
Future Perspectives

Abstract Water cycle management will direct the design of future urban water and wastewater system in a new manner that the reclaimed water can be produced and utilized most efficiently and the demand on freshwater resource can be reduced to the minimum. For achieving this goal, the methods for biological and human health safety control still need to be further studied.

Keywords Water cycle management · Water application efficiency · Safety control

Water cycle management is, in fact, not a new terminology in the hydrological field because, as discussed in Chap. 2 of this book, human beings have depended on the global and basin-scale hydrological cycles for obtaining freshwater for daily life and economic activities since ancient time (Cech 2005). However, the introduction of the concept of water cycle management into urban and/or district water systems, especially with consideration on water reuse only began recently under the condition of insufficient water resources in many countries and regions to meet the needs for population increase and urban development (Tambo et al. 2012). If we recall the history of human utilization of water resources, it experienced the early stage of scattered water intake directly from natural waters for settlements and villages, and the next stage of water supply by aqueducts for ancient cities of small scales. Used water discharge or disposal was also done in completely natural ways at that time through direct discharge to natural waters. However, such kind of extremely simple methods for water and sanitation did not result in any noticeable contamination because the scale of water utilization was very small comparing with the capacity of self-purification of natural waters (Vagnetti et al. 2003). The turning point to what we call the modern urban water systems was at the 18th century as a result of the industrial revolution, which brought about the birth of modern cities in Europe and then America and other continents. For urban water supply, provision of pipeline networks covering the whole service area is the first feature of the modern urban system. Regarding water quality control, it underwent direct supply, slow sand filtration and ultimately the rapid sand filtration process along with the important inventions of alum as a coagulant and chlorine for disinfection

X.C. Wang et al., *Water Cycle Management*,
SpringerBriefs in Water Science and Technology,
DOI 10.1007/978-3-662-45821-1_5

(Pontius and Clark 1999). The modern sewerage system was also formed in the same period with the provision of piped sewers and the rapid development of wastewater treatment technologies from land treatment, chemical treatment, to the activated sludge process (Cooper 2001). The driving force for the technological development was from the requirement for better sanitary conditions in growing cities, and the protection of natural waters.

As the modern urban water and wastewater system aimed at provision of high quality water with sufficient quantity to meet the demands for various water consumption, and discharge of human wastes as completely and swiftly as possible out of the urban area, it was characterized by water use and discharge in the "end-of-the-pipe" manner, namely using large quantity of water not only for portable purposes, but also for "washing out" wastes and discharge all the used water to a receiving water body (Wilderer 2001). It should be pointed out that such a manner of water utilization was developed in a time when natural water was thought to be plentiful, and water saving or water reuse was not an issue to be considered. On the other hand, the modern urban water and wastewater system connects with the natural waters only at water intake where source water is withdrawn and the outlet of urban drainage where used water is received by the natural water. The system itself is virtually an engineered pathway for water to flow, with several stages of water quality conversion such as drinking water purification and wastewater treatment. We can say that from ancient time to nowadays, the way of water utilization has gradually evolved from natural dependence to engineering dependence. One of the important reasons for this is the rapid increase of ecological footprint due to population increase and economic development. It is estimated that in 1970 the global ecological footprint began to exceed the global biocapacity, and in 2010 the former was already 1.5 times of the latter, indicating that we need one and a half planet Earths to accommodate human beings (WWF 2014). Under such conditions, it is emergent for us to consider from now how to save the very limited natural biocapacity. One thing directly related to this is to change the way we built the urban water and wastewater systems in the conventional manner because it is no longer suitable to the current situation of limited available resources and biocapacity.

The objective of water cycle management of the water reuse system discussed in this book can be explained as to build an almost biocapacity-self-sufficient water district where the pollutants discharged to the environment outside the district is minimum so that the biocapacity required to assimilate the extra pollutants is saved. On the other hand, through the water cycle, as shown in the case discussed in Chap. 4, the efficiency of water utilization is almost doubled. This in due saves the biocapacity for additional natural water resource development. The water area, such as an artificial lake created using the reclaimed water also contributes to the increase of the biocapacity in the district.

To facilitate water reuse, decentralized systems are recognized as potential alternatives of the conventional centralized systems (Chen and Wang 2009; Mankad and Tapsuwan 2011; Marlow et al. 2013). The difference between the centralized and decentralized systems is not in the system scale, but in the efficiency of water reclamation and impacts on the water environment. Every watershed has its limited capacity to serve human beings and, meanwhile, to maintain itself at a healthy condition. The system to be planned or designed should be under the restriction of the local environmental capacity. Following such a principle, the system can be designed as larger as possible and as smaller as necessary, but the principle of water cycle management may always need to be followed.

For the reclaimed water to be used through a water cycle, its quality and safety are a matter of concern. In addition to individual water quality parameters, which are specified in many standards regarding various water uses, the ecological safety and pathogenic safety are important topics of discussion in this book. Regarding the ecological safety, attention has been paid to the impacts of chemical pollutants residual in the reclaimed water on aquatic organisms and the main pathway of exposure is recreational impoundment. In comparison with the chemical analysis to determine the components and concentrations of the pollutants in water, biological analysis can intuitively reflect the detrimental effects of pollutants on organisms and profoundly reveal the mechanisms of the toxic effects including formation, development and removal (Li et al. 2013a, b). This leads to the development of bioassays using sensitive and observable organisms including specific species of bacterium, alga, daphnia, protozoan, and fish. Of these organisms, because luminescent bacteria are with short reproduction cycles, easy to be cultivated and to be observed, the luminescent bacteria inhibition assay has become the first choice for ecotoxicity assessment (Ma et al. 2014). The bioassays using luminescent bacteria can be used for both the comprehensive assessment of the toxicity due to all toxic substances in water and the toxicity due to specific toxins. Moreover, based on experimental data, establishing the quantitative structure-activity relationships of chemicals and analyzing the joint effects of mixtures can help to describe the mechanisms of chemicals to luminescent bacteria, to obtain the toxicity results of untested samples and to determine the toxicity interaction of mixtures (Ma et al. 2014).

The pathogenic safety of the reclaimed water is closely related to human health protection. Unlike chemicals, pathogens may cause acute infections with a very low dose upon exposure, even in the case of non-potable reuse of the reclaimed water (Asano et al. 2007). For a comprehensive assessment of human health risk, the conventional fecal indicator such as fecal coliform cannot provide sufficient information. Therefore, quantitative detection of specific pathogenic bacteria and viruses becomes necessary, and this leads to the development and/or application of molecular biological techniques such as PCR and RT-PCR discussed in Sect. 3.3.1. 2. Besides the quantitative detection of specific pathogens, these techniques in combination with other methods can assist the identification of virus sources for rare diseases such as Hand-Foot-and-Mouth Disease (Ji et al. 2012), and the analyses of genotypic diversity of enteroviruses from different wastewaters (Ji et al. 2014). As long as the distribution of infectious pathogens is understood, human health risk can be quantified following the dose-response relationships. However, because disease infection may occur in a single case of exposure to pathogens, pathogenic safety control should be based not only on the annual average risk control but also the severe single case risk control. The latter may be more important in the case of reclaimed water use.

References

Asano T, Burton FL, Leverenz HL, Tsuchihashi R, Tchobanoglous G (2007) Water reuse: issues, technologies, and applications. Metcalf & Eddy, McGraw-Hill, New York

Cech TV (2005) Principles of water resources—history, development, management, and policy. Wiley, Hoboken

Chen R, Wang XC (2009) Cost–benefit evaluation of a decentralized water system for wastewater reuse and environmental protection. Water Sci Technol 59(8):1515–1522. doi:10.2166/wst.2009.156

Cooper PF (2001) Historical aspects of wastewater treatment. In: Lens P, Zeeman G, Lettinga G (eds) Decentralized sanitation and reuse: concepts, systems and implementation. IWA Publishing, London, pp 11–38

Ji Z, Wang XC, Zhang CM, Miura T, Sano D, Funamizu N, Okabe S (2012) Occurrence of hand-foot-and-mouth disease pathogens in domestic sewage and secondary effluent in Xi'an, China. Microbes Environ 27(3):288–292. doi:10.1264/jsme2.ME11352

Ji Z, Wang XC, Xu LM, Zhang CM, Funamizu N, Okabe S, Sano D (2014) Estimation of contamination sources of human enteroviruses in a wastewater treatment and reclamation system by PCR–DGGE. Food Environ Virol 6:99–109. doi:10.1007/s12560-014-9140-x

Li S, Anderson TA, Maul JD, Shrestha B, Green MJ, Cañas-Carrell JE (2013a) Comparative studies of multi-walled carbon nanotubes (MWNTs) and octadecyl (C18) as sorbents in passive sampling devices for biomimetic uptake of polycyclic aromatic hydrocarbons (PAHs) from soils. Sci Total Environ 461–462:560–567. doi:10.1016/j.scitotenv.2013.05.048

Li S, Irin F, Atore FO, Green MJ, Cañas-Carrell JE (2013b) Determination of multi-walled carbon nanotube bioaccumulation in earthworms measured by a microwave-based detection technique. Sci Total Environ 445–446:9–13. doi:10.1016/j.scitotenv.2012.12.037

Ma XY, Wang XC, Ngo HH, Guo W, Wu MN, Wang N (2014) Bioassay based luminescent bacteria: interferences, improvements, and applications. Sci Total Environ 468–469:1–11. doi:10.1016/j.scitotenv.2013.08.028

Mankad A, Tapsuwan S (2011) Review of socio-economic drivers of community acceptance and adoption of decentralised water systems. J Environ Manage 92(3):380–391. doi:10.1016/j.jenvman.2010.10.037

Marlow DR, Moglia M, Cook S, Beale DJ (2013) Towards sustainable urban water management: acritical reassessment. Water Res 47(20):7150–7161. doi:10.1016/j.watres.2013.07.046

Pontius FW, Clark SW (1999) Drinking water quality standards, regulations, and goals. In: Letterman RD (ed) Water quality and treatment—a handbook of community water supplies, 5th edn. McGraw-Hill Inc., New York, pp 1.1–1.45

Tambo N, Nakamura M, Wang XC (2012) The fractal lentic-lotic basin complex as a conceptual basis for future urban water systems. J Water Sustain 2(2):105–120. doi:10.11912/jws.2.2.105-120

Vagnetti R, Miana P, Fabris M, Pavoni B (2003) Self-purification ability of a resurgence stream. Chemosphere 52:1781–1795. doi:10.1016/S0045-6535(03)00445-4

Wilderer PA (2001) Decentralized versus centralized wastewater management. In: Lens P, Zeeman G, Lettinga G (eds) Decentralized sanitation and reuse: concepts, systems and implementation. IWA Publishing, London, pp 39–54

WWF (2014) Living planet report 2014. World Wildlife Fund, Gland

MIX
Papier aus verantwortungsvollen Quellen
Paper from responsible sources
FSC® C105338
FSC
www.fsc.org

If you have any concerns about our products,
you can contact us on
ProductSafety@springernature.com

In case Publisher is established outside the EU,
the EU authorized representative is:
Springer Nature Customer Service Center GmbH
Europaplatz 3, 69115 Heidelberg, Germany

Printed by Libri Plureos GmbH
in Hamburg, Germany